TIGER TALES

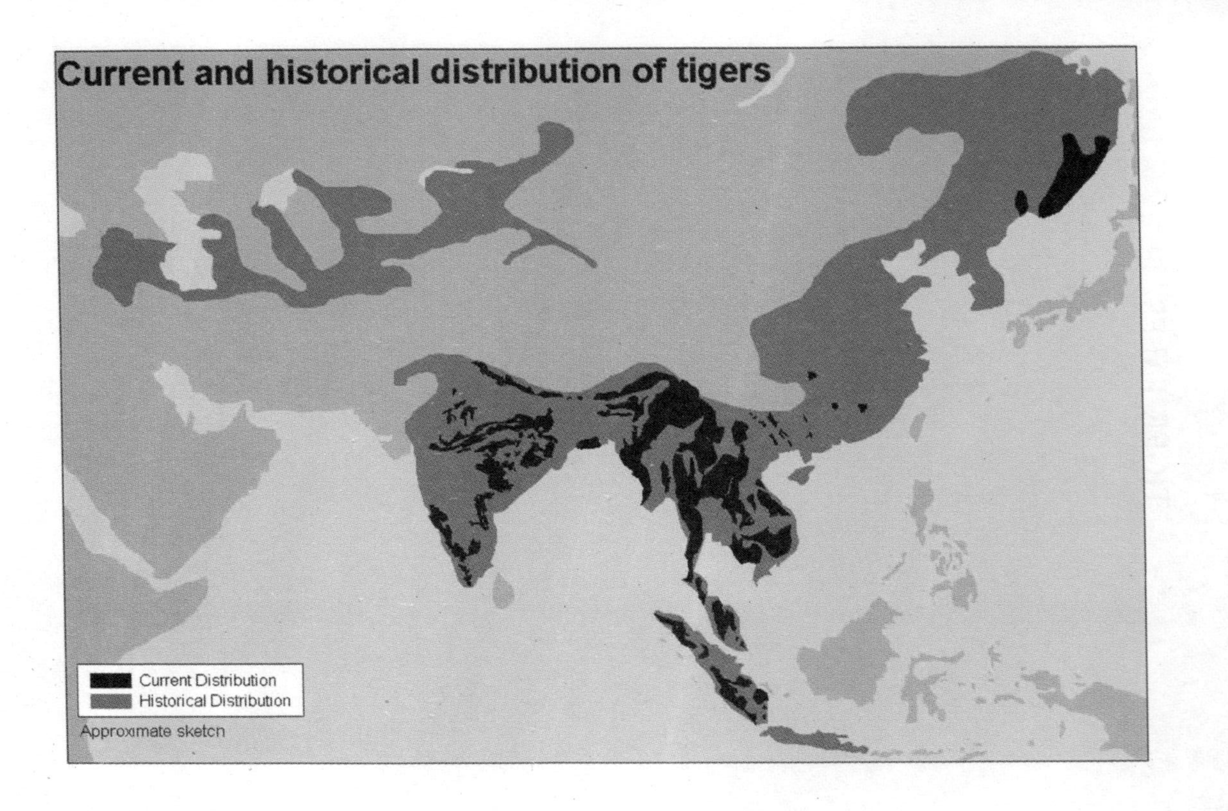
Current and historical distribution of tigers
Current Distribution
Historical Distribution
Approximate sketcn

TIGER TALES

Tracking the Big Cat across Asia

EDITED BY
K. ULLAS KARANTH

PENGUIN BOOKS

PENGUIN BOOKS
Published by the Penguin Group
Penguin Books India Pvt Ltd, 11 Community Centre, Panchsheel Park, New Delhi 110 017, India
Penguin Group (USA) Inc., 375 Hudson Street, New York, New York 10014, USA
Penguin Group (Canada), 90 Eglinton Avenue East, Suite 700, Toronto, Ontario, M4P 2Y3, Canada (a division of Pearson Penguin Canada Inc.)
Penguin Books Ltd, 80 Strand, London WC2R 0RL, England
Penguin Ireland, 25 St Stephen's Green, Dublin 2, Ireland (a division of Penguin Books Ltd)
Penguin Group (Australia), 250 Camberwell Road, Camberwell, Victoria 3124, Australia (a division of Pearson Australia Group Pty Ltd)
Penguin Group (NZ), cnr Airborne and Rosedale Roads, Albany, Auckland 1310, New Zealand (a division of Pearson New Zealand Ltd)
Penguin Group (South Africa) (Pty) Ltd, 24 Sturdee Avenue, Rosebank, Johannesburg 2196, South Africa

Penguin Books Ltd, Registered Offices: 80 Strand, London WC2R 0RL, England

First published by Penguin Books India 2006

Page 297 is an extension of the copyright page

10 9 8 7 6 5 4 3 2 1

ISBN-13: 9780144001385 ISBN-10: 0144 001381

Typeset in Sabon by Mantra Virtual Services, New Delhi
Printed at Baba Barkhanath Printers, New Delhi

This anthology is dedicated to two fine foresters from India who mentored me over the years and nurtured my interest in wild tigers

S. SHYAM SUNDER

and

K. M. CHINNAPPA

CONTENTS

ACKNOWLEDGEMENTS

I thank the following contributing writers who I know personally, for allowing the use of their fine articles: George Schaller, Melvin Sunquist, Fiona Sunquist, John Seidensticker, Dale Miquelle, John Goodrich, Linda Kerley, Mahesh Rangarajan, Geoffrey Ward, Valmik Thapar, Billy Arjan Singh and Anne Wright. I am also indebted to all the other contributors—some of whom are no more—whose writings I have enjoyed greatly.

I am grateful to Debbie Behler, Steve Johnson and Kae Kawanishi for help in locating some articles; to Samba Kumar and Devcharan Jathanna for preparing the map of tiger distribution; and Maya Ramaswamy for the illustration. I thank the Wildlife Conservation Society, New York, for having supported my work over the years.

I would like to acknowledge V.K. Karthika and Paromita Mohanchandra at Penguin Books India who convinced me that this effort was a worthwhile undertaking and, thereafter, encouraged me in numerous ways through their editing skills, savvy and inordinate patience.

K. Ullas Karanth

INTRODUCTION

A JOURNEY INTO THE TIGER'S WORLD

Tigers have inspired people the world over—across time, space and cultures. This fascination for the tiger has generated a veritable flood of literature in the form of fiction, folklore, myths, shikar tales, natural history and science. My task was to select a few articles that provide readers with a key to open the doors to this enormous tiger library. To make this task a little more tractable, I made some rules. Some of these rules were objective, while others unabashedly subjective. Before delving into the contents of the anthology that resulted, I will try to explain the basis for these choices.

In the public mind, tigers are inextricably linked with India. Yet, wild tigers had once roamed about thirty of the present-day nation states, which stretch all the way from Armenia to Indonesia, from Russia to Iraq. Although the majority of surviving wild tigers are assumed to live in India now, only 20 per cent of the remaining tiger habitat occurs here. Although India is assumed to hold half the world's tigers, the rest of Asia has three-fourths of the remaining habitat. Across the tiger's range, the animal's ecology as well as the human cultures that imperil it, both vary greatly. Therefore, I have devoted roughly a quarter of this book to tigers outside India.

As a professional conservation scientist, I feel distinctly unqualified to assess the rich fare of fiction, myths and folklore about tigers that exists. So I decided to leave tiger fiction and tiger lore out, although compromising somewhat by including some hunting tales that almost border this domain. Even within the limits of 'tiger non-fiction', I was forced to make difficult choices from among excellent articles, which dealt with myriad tiger issues. I did not choose stories about captive tigers because I find them less relevant. I hope the readers share my biases.

I have somewhat arbitrarily pigeonholed the varied tiger topics covered by the authors under three broad themes: tiger hunting and old natural history, tiger preservation and new natural history, and, modern tiger science and conservation. Within each category, the essays presented follow an approximately chronological sequence. I primarily chose writings from the late-colonial and post-colonial periods when the very survival of wild tigers came under threat, and dramatic recovery efforts ensued to save the big cat.

Most of these essays are popular or semi-popular articles. They represent two writing styles: tales that convey the immediacy, excitement and details of tiger incidents experienced by the authors, and more reflective analyses of tiger issues, based on either reason or polemics. The authors I present here are all naturalists of one kind or other, but they looked at tigers from different perspectives: as hunters, soldiers, administrators, foresters, game managers, conservationists, scientists, historians and so forth. I had to balance the cumulative product in terms of topics, geography and diversity of perspectives, without compromising narrative flow. For one or the other reason, I had to exclude some fine articles that I had liked. I also had to delete parts of selected articles, although I did not significantly amend any of the original text that I chose to retain.

This exploration of the tiger's world begins with Mahesh Rangarajan's thoughtful analysis (from his *India's Wildlife History*, 2001) of how Indian tigers fared under colonial and princely rulers in the post-Mughal era. While Rangarajan grudgingly credits these ruling classes for establishing some protected game reserves, he also clinically exposes the horrendous scale of the tiger slaughter indulged in by the social elite either through sport hunting, or through bounty hunting they encouraged among subaltern classes. This analysis provides the necessary background for absorbing the first-hand tiger experiences that follow.

Many of us have *a* favourite piece of tiger turf. Mine is the 1500 square kilometre stretch of forest covering Bandipur and Nagarahole reserves in Karnataka. This is where I learned my jungle craft, and later studied my tigers. The colonial forester C.E.M. Russell lorded over this tract between 1882 and 1896. The account presented here is drawn from his *Bullet and Shot in Indian Forest, Plain and Hill* (1900). Russell offers some blunt advice to wannabe hunters about the intricacies of 'beating' the jungles for tigers, and on conscripting the reluctant 'natives' into this dangerous venture. Russell's tale of following a wounded tiger in the Naganapura forests (now a part of Bandipur) provides a hair-raising read. The author laments the scarcity of tigers in the Mysore region as early as in the late nineteenth century. He attributes their decline to the bounty hunting and poaching of deer and pigs by local people who, he complains, were encouraged by an 'extremely lenient government' and 'a rabid, scurrilous native press'.

Among colonial forester-naturalists who followed Russell, A.A.D. Brander stands tall. Unlike his colleagues who only tried to kill tigers, Brander had almost stopped shooting for six years to concentrate on watching these big cats in the central Indian forests that now include the tiger reserves of Kanha, Pench, Tadoba and Melghat. Presented here are Brander's careful observations on the size, colouration and predatory behaviour of tigers drawn from his classic work *Wild Animals in Central India* (1923).

Another forester, F.W. Champion describes in *With a Camera in Tiger Land* (1927) how he pioneered tiger photography using trip-wire activated cameras in the early twentieth century. Champion set his primitive 'camera traps' near tiger kills and along game trails in the Himalayan foothills. He got a picture only *if* a tiger came along, and *if* it tripped the wire correctly, and *if* the wire successfully activated *both* the powder-flash and the camera shutter. Although Champion set around 200 such traps during his entire career, tigers came by them only on eighteen occasions. Moreover, Champion's clunky apparatus worked only eleven times, to reward him with just nine decent tiger pictures for a lifetime of effort. Every year, scientists like me now deploy dozens of modern electronic camera traps and get hundreds of tiger pictures for research. Nonetheless, Champion's pioneering spirit and his stunning tiger images never cease to amaze me.

Jim Corbett's writings ignited the spark of natural history in several generations of Indians and Corbett's description of the slaying of the dreaded Kanda man-eater in his *Man-eaters of Kumaon* is one of my favourite tales. However, due to problems in obtaining permission to reprint any of his writings I was unable to include him in this anthology. Kenneth Anderson's tale from *This is the Jungle* (1964) is also cast in a Corbett-like mould. Although Anderson was an even more gifted storyteller than Corbett, I have a suspicion that his man-eating tigers straddle the world between fact and fiction. I got to know Anderson rather late in his life, but still cherish fond memories of following tiger spoor with him in the locale where this story is set.

Tigers originally evolved two million years ago in eastern Asia as specialized predators of large herbivores, moving into the Indian subcontinent only a few thousand years ago. However, because of our collective obsession with Indian tigers, we tend to ignore tigers in the rest of the world. Yet, early in the last century, the vast savannah-woodlands of Indochina teemed with tigers. These forests probably equalled the best of tiger habitats in India, because they supported teeming herds of four mega-herbivore species: wild water buffalo, banteng, gaur and kouprey. Tragically, hunters extirpated the kouprey soon after it was discovered by modern science in 1938. They also greatly depressed numbers of all other prey species, thereby pushing tigers to the brink of extinction within a few decades. French colonist William Bazé hunted big game in Indochina until 1945, and recorded his exploits in the book *Tiger! Tiger!* (1957). Bazé vividly describes the great faunal wealth of Indochina, as well as some of the unusual methods men like him employed to bankrupt it so thoroughly.

The perennially damp tropical rainforests, where tall evergreen trees block out most of the sunlight, are not the best of tiger habitats. Large-hoofed prey animals that support tiger populations tend to be scarce in these forests. However, tigers do survive in many tropical rainforests of Southeast Asia. They tend to be smaller, and their ecology somewhat different in comparison to their cousins in India and Russia.

Beginning in 1949, Arthur Locke administered the rainforest-rich Trengannu State in Malaya for only two years. However, Locke's shooting skills must have attained his self-imposed proficiency standard of 'hitting an un-husked coconut, three times in succession, at 20 paces' rather quickly, because he destroyed several man-eaters and cattle killers during this short tenure. However, like Jim Corbett before him, Locke greatly admired the richly striped cats he was compelled to kill in the line of duty. Locke's observations on the food and movements of rainforest tigers drawn from his *Tigers of Trengannu* (1954) occupy a special niche in the realm of tiger natural history accounts.

The first few decades that followed the departure of colonial powers from Asia, turned out to be a tragic period for tigers. Even in Jawaharlal Nehru's 'enlightened' India, vast stretches of forests were logged and cleared, prey animals shot wholesale, and the tigers themselves persecuted viciously—all in one big wave of 'progress' unleashed in the hope of uplifting the newly free but desperately poor masses. During this era, side by side with old-style sport hunting, the menace of African-style commercial hunting also raised its ugly head. I still recall with distaste how assorted 'sportsmen'—some of whom even managed to re-label themselves as conservation pioneers later—ignored the all too obvious warning signs and continued killing tigers well into the 1960s. The essays by Kesri Singh, a game manager under Indian princes (*The Tiger of Rajasthan*, 1959), and Jack Denton Scott, an American safari hunter (*Forests of the Night*, 1959) provide us with grim examples of the 'sportsmanship' that prevailed in this dismal period.

In the next section, having had our fill of tiger hunting and natural history observations made along the sights of a rifle, we step into the world of tiger preservation and the new natural history that was driven by the sheer joy of watching the animal.

A. Hoogerwerf's book *Udjung Kulon: The Land of the Last Javan Rhinoceros* (1970) is set in a wet forest reserve of Indonesia. Excerpts here illustrate not just a change of scene, but also the gradual shift in societal attitude away from killing of tigers towards saving them.

Hoogerwerf was a colonial conservationist who managed Udjung Kulon, beginning in 1938. He provides us with first-hand observations of tigers sliding down to extinction here. This Dutch naturalist was the only man ever to photograph a wild Javan tiger. Ironically, the rhinos that Hoogerwerf worried about have survived in Udjung Kulon whereas the tigers he wrote about with such understated poignancy disappeared soon after.

Moving on, we enter the world of tiger preservation through the tea planter turned conservationist E.P. Gee's article, extracted from *Wildlife of India* (1964). This book was probably the first one to arouse middle-class India's conscience to the post-independence decline of wildlife. It was the intellectual trigger for the strong conservation initiatives that emerged subsequently. Gee's account also sets the stage for the next article: Guy Mountfort's reminiscences (*Saving the Tiger*, 1982) about how a small coterie of international conservationists tried to convince Indian Prime Minister Indira Gandhi and other Asian leaders about the urgency of saving the big cat.

Kailash Sankhala (*Tiger!*, 1978) takes up the story from where Mountfort stops. He describes the political and administrative strategies that underpinned efforts of Indian foresters in nurturing the nascent Project Tiger initiative. Sankhala lauds the project's ecosystem approach, its mission focus, and the managerial abilities of his field team. He goes on to cautiously proclaim the project a success. In the next essay, however, conservationist Arjan Singh (*Tiger! Tiger!*, 1984) takes a contrary view through the sceptical eyes of a seasoned tiger-watcher. Singh argues that India's official tiger conservation efforts are bogged down by incompetence, corruption and ecological ignorance. He pleads the case for creating a brand-new professional wildlife service for managing India's tigers.

Breaking off from this polarized debate, we look at the 'new natural history' that blossomed after tiger populations rebounded in some protected reserves. In places like Ranthambhore, tigers habituated to tourist presence could be watched in broad daylight, like never before in history. Valmik Thapar's essay on tiger watching (*The Secret Life of Tigers*, 1989) brings the big cat's daily life alive before us.

In Indian tiger reserves like Ranthambhore, Kanha and Nagarahole, strict protection led to increased prey abundance, allowing tiger populations to attain high densities. The forests of the Russian Far East, however, represent the other natural extremity of the tiger's ecological adaptation. In the temperate Taiga, prey like deer and pigs are naturally scarce, and, during the sub-zero winters, tigers can lose their toes to frostbite. Russian naturalist Vladimir Troinin's account (1994) tells us how big cats—and conservationists—survive in this harsh 'tiger-eat dog' world.

Despite some conservation successes of the 1970s and 1980s, the tiger's struggle to survive will continue into the foreseeable future. Modern wildlife science will be the keystone in our efforts to prevent the big cat's extinction.

In the concluding section of the book, we switch from subjective natural history to objective conservation science and the reliable knowledge the latter generates. These accounts begin with George Schaller's classic report on the tigers of Kanha, which first appeared in *Life* magazine in 1965. Schaller's was the first-ever tiger study conducted in the framework of modern wildlife biology. *The Deer and the Tiger*, a book that Schaller published two years later, is perhaps the single most important tiger book ever written.

Although Schaller used only a pair of binoculars and a notebook to study tigers, his work inspired a new generation of biologists to probe deeper into the tiger's secrets employing more advanced tools. First among these biologists was John Seidensticker, the lead author of the article on problem tigers of the Sundarbans (*Oryx*, 1976), which clearly illustrates the ecological and social issues that must be faced when dealing with human–tiger conflict. Seidensticker initiated the first ever radio-telemetry study of wild tigers in Nepal in the early 1970s, a project that was brilliantly executed by his successors Mel Sunquist and Dave Smith in association with Chuck McDougal. The Chitwan study continued over a decade and is recognized as a landmark in tiger conservation. Here, Sunquist and his naturalist wife Fiona summarize the key findings of their research on the social life of tigers (*Natural History*, 1983).

The next example, reproduced from *Wildlife Conservation* (1995) draws on my own research work involving the capture, tracking and camera trapping of wild tigers in the forests of Nagarahole. This study, now into its twentieth year, continues to generate new insights on how tiger populations function across large landscapes. Another tiger-science essay is also drawn from *Wildlife Conservation* magazine (2002). It is co-authored by a team of biologists (John Goodrich, Dale Miquelle, Linda Kerley and Evygeny Smirnov) who have been studying tigers in the Russian Taiga for over a decade now. Their account precisely illustrates how tiger science can inform and aid tiger conservation practices.

The concluding piece of this collection is a popular *tour de force* of global tiger conservation efforts by historian Geoffrey Ward that was originally published in the *National Geographic* magazine (1997). I believe, as a lucid summary of tiger issues in our changing world, this essay is hard to beat.

Exploring the splendid world of tigers has given me great pleasure and satisfaction over the years. I hope that this anthology will encourage the readers also to venture more deeply into that fascinating world.

K. Ullas Karanth

PART 1

HUNTING AND OLD NATURAL HISTORY

MAHESH RANGARAJAN

VENOMOUS SNAKES AND DANGEROUS BEASTS

Even by the standards of English eccentricity, Major Tweedie's suggestion in 1874 surely borders on the bizarre. In a crucial debate on how to control dangerous wild animals in India, he advised:

> It becomes a question how far it would not be well to employ in each region where necessity exists a certain number of paid tiger-killers or snake-destroyers, as the case might be, whose sole and special duty would be to follow their vocation just like the mole-catcher and rabbit-killer in our country. If the extermination of creatures which prey upon herbivores were taken up as systematically in India as the extermination of creatures which prey on game in England, there is no reason why very satisfactory results should not soon be obtained.[1]

Though the idea was not taken up except in a handful of districts (with very mixed results), bounties were given out in various provinces to eliminate 'dangerous beasts and poisonous snakes'. The village-based shikari, often a low caste hunter, became the lynchpin of the Raj's efforts to impose on South Asia's jungles its own vision of nature. Eventually, a centralised administrative machine began to oversee such efforts, resulting in a veritable war against errant species. They were a scourge to be wiped out. Such practices were new to India: no previous ruler had ever attempted to exterminate any species.

Within two decades of defeating the rulers of Bengal in the historic Battle of Palashi in 1757, the British decreed special rewards for any tiger killed. Other legitimate targets included large herbivores like the elephant, the wild buffalo and the great Indian one-horned rhinoceros. There is no doubt there were points of conflict between mega-mammals and people before the coming of European rule but these now acquired a sharper edge than ever before. Rulers who preceded the British had often asked their local officials to eliminate tigers, bandits and thieves. The idea was to help push back the jungle. It was part of the constant tug of war between axe and plough on the one hand and the incredible ability of natural vegetation to spring back. The tug of war now became a fight to the finish.

Part of the British animosity to the forest and its wild inhabitants stemmed from the situation in Bengal, the very first region of India they conquered. Much of eastern India suffered a major famine around 1770, with one in three people dying. The excessive revenue demands made by the new colonial rulers was a major culprit. As a result of the massive mortality, large areas of farmland remained uncultivated and reverted to jungle. The secondary growth was probably ideal habitat for deer and wild boar and their chief predator, the tiger. Bounties aimed to eliminate cattle-marauding tigers. Saving draught cattle would help extend the area that was under the plough. Fewer tigers meant more cultivation and more revenue, their elimination a blessing of imperium after the elimination of an oriental despot. Unprecedentedly, larger rewards were given out for killing tigresses, and special prizes for finishing off cubs. This was to be a war where no quarter was given. It was not only an issue of self-interest, important as the land revenues of Bengal were in subsidizing military conquest and trade elsewhere. The image of the tiger was also imprinted in the official British mind as a flesh-eater that dared to eat *people*. Similar techniques had been honed to perfection in the British Isles whose prime predator, the wolf, had already been killed off by the time the British founded an empire in India.

The sport hunters who shot the tiger in huge numbers saw it in a largely negative light. It was 'a cunning, silent, savage enemy', a pleasure to outwit and shoot, to end 'the fearful ravages' it committed against the people. This kind of propaganda whetted the blood-thirst of generations of army officers, civil officials and box-wallas (as the traders were known). District level administration went out of its way and facilitated hunts, local landed gentry lent their elephants, and peasants were pressed into action as beaters. Officers and soldiers in cantonments were encouraged to spend their vacations acquiring more trophies. Stories of their hunts for wild prey figured in after-dinner anecdote alongside tales of military campaigns. The tiger was the prime example of a lawless beast, whose conquest was held to be among the greatest blessings conferred by Pax Britannica. The relative shyness of the animal was read as a sign of its essentially depraved nature and eliminating large wild mammals, including tigers, made the jungle safe for the woodcutter and cattle-grazer. Protective hunts were deeply symbolic of the logic and rhetoric of empire, of brave white men defending hapless mothers whose children fell prey to wild beasts. As T.T. Cooper, a big game hunter in Assam said of the wild buffalo, it was 'so numerous and so destructive as to be an absolute pest'.

Of course, it followed that the new rulers, with their massive interventions, unleashed new conflicts that led to increased fatalities in the encounters between people and carnivores. Initially the bounties had the desired effect in large parts of Bengal Presidency, bringing down the loss of human lives and killing off large numbers of target species. In the medium term things took a turn for the worse. Many animals classed as vermin were probably concentrated in greater numbers along the edge of grasslands and mature tree forest. The growth of plantation crops such as tea in the hills of Assam reduced the available habitat and, for a time, increased the intensity of conflict. As the agrarian frontier extended, the concomitant growth in human numbers multiplied the chance of deadly encounters on the ground. Conversely, the slaughter of deer and boar by sahibs or villagers out to get extra meat reduced the base of prey for wild carnivores. The rhino and wild buffalo, major prey items, vanished from the north Bengal plains by the 1850s; in the drier regions, the nilgai became scarce. The shifts in faunal distribution had their counterpart in vegetational changes. The greatest change was that the entire landscape was being divided permanently into forest and farm. A continuum of tree forest, savannah and abandoned farmland was giving way to a countryside divided into two landscapes, of cultivated space or of forest.

The 1850s were a time of political turmoil: the rebellion of the Santhals in the east was followed by the Great Rebellion (earlier labelled the Mutiny) of 1857. One immediate consequence was the crackdown on anything that could lead to 'disorder'. Colonial strictures against the annual hunts of the Santhal tribals removed a major check on wild animal populations. The disarming of peasants in the aftermath of the Rebellion of 1857–8 often deprived them of the means of effective self-defence against wildlife. Various local systems of control and self-defence were being replaced by a new regime that sought to resolve conclusively the issue of human–wildlife relations. Above all, tigers injured by sport hunters themselves became a menace to human life and property. It is quite possible hundreds or even thousands of wild creatures were injured, having been shot at but not killed outright. The common refrain in the old hunting manuals was that a true sportsman ought to follow up and finish off a wounded predator. The fact was that this injunction was easily disregarded. Tigers wounded by gunshot turned to preying on livestock, and more rarely, on people.[2]

In the latter half of the nineteenth century, the drama in Bengal was re-enacted on a subcontinental scale. In the 1870s, local practices across British-ruled territories were evaluated and the Government of India worked hard to assess the best method of exterminating wild animals. By the time Tweedie wrote his note, over 20,000 animals were being killed for bounties in British India each year. In turn, about as many people were reported killed by various creatures including carnivores and poisonous snakes. Snakes claimed four out of every five human lives lost. Even today, several thousand lives are lost to snakebite every year due to the lack of timely medicinal remedy. This also means most snakes are great survivors and can inhabit the spaces occupied by people. Figures apart, there is little doubt that there was a new kind of tension in the human relationship with the larger denizens of the forest. Among the creatures targeted for extermination was the cheetah, even though it was largely a harmless animal. At the southern edge of its range, in Madras Presidency, Collectors paid out twenty-five rupees for each specimen. Though not a prestigious sporting trophy, it was targeted as vermin. At imperial urging, many princes who ruled a third of India's landmass also gave out rewards. The Maharao of Kotah in Rajasthan awarded anyone who killed a lion twenty-five rupees, more than twice what he gave for the head of a tiger. In the North West Provinces, the wolf got its killer a bounty larger than that paid for a panther. Wolves were a larger threat to stock and, in some districts, to young children.

The choice of target was based on a mix of economic and cultural factors. Carnivores were not taken off the lists of vermin, unless they were important for sport, though this happened in princely states long before it did in British India. The elephant, important for warfare and carriage, was removed from the list of vermin and given legal protection in southern India. The wolf was listed as a pest till well after Independence.[3]

Responses to the rewards were mixed. One reason was the great diversity of situations that could exist even within one province or district. Often, the attitudes of people living on the forest edge determined the degree to which vermin extermination was a success or a failure. There is no doubt it had some support. The extermination of the arna or the great wild buffalo often found enthusiastic support among rice-growers and cattle-herders in the wet savannahs along the Brahmaputra and the Ganga rivers. It vanished even in areas with substantial habitat because of over-enthusiastic bounty hunters and sportsmen. Similarly, the wolf was an inveterate foe of goat-keepers and shepherds who inhabited dry rainfall zones in the Deccan and in north India.

But animosities could have their limits. Cultivators and even sections of the landed gentry saw in the tiger an ally who preyed on herds of deer, and a sounder of wild boar. Unlike parts of Bengal, there were large regions in southern India where man-eating was virtually unheard of: officials admitted that the striped cat was entitled to 'present his little account for services rendered in keeping down wild animals which destroy crops'. Even the wild buffalo had its defenders in areas where the owners of herds of domestic buffalo wanted them to interbreed and produce strong male offspring.

Earlier local equations had determined the shape of conflict between the human land-user and wild animals. They still mattered, but material rewards were a potent new factor in the situation. At times, they did tilt the scales in conflicts. The sheer ingenuity of the peasant, however, sometimes defeated the grand designs of officials. Even those with little love for carnivores soon found ways to manipulate the system of bounties, sometimes defeating its very purpose. They supplied fake skins, or heads of animals long dead, simply to get their share of money. Officials were ordered to carefully check on the skins before paying out any money from the treasury. Government personnel often did not know the difference between a jackal and a wolf. Their ignorance was even more profound when the carcass in question was that of a cub. The peasant soon worked this out. Nomadic hunting tribes collected the heftier bounty on the wolf killing only the smaller, relatively harmless jackal.

Religious and cultural values could also act as brakes on outright slaughter. In Salem district near Madras, shepherds hesitated to kill either wolves or wild dogs. Some even believed that if a wolf lifted one goat from a flock, the rest of the herd would thrive and do well. Such views could vary enormously from one region to another or between and across different ethnic groups. In some parts of its range, the tiger was viewed with almost religious awe, with residents hesitant to even take its name lest it do them harm. Even outspoken champions of tiger extermination conceded that villagers often viewed the great cat with 'remarkable indifference'. It is probable that there was a fair degree of tolerance or coexistence in forest tracts with plentiful wild prey.[4]

There were voices raised in favour of letting certain increasingly rare species off the hook but with little impact. Even this was a major shift in sensibility: a generation earlier, such extermination would merely be seen as a sign of progress. In 1907, the veteran sportsman Reginald Gilbert was urging that the general bounty on tigers be suspended. Only known man-eaters ought to be killed, he said; the decline of fauna in the Indian Empire required urgent remedies. The tiger found a few friends partly due to its sheer prominence as a trophy-worthy species. There were fears that its disappearance would mean the end of the finest of sports India had to offer its rulers. This eased but did not end the pressure: there were still several powerful people who felt a tiger was better dead than alive. Old images died hard. But over time, the more lurid versions of its ferocity gave way to a realization that most tigers simply left people alone. By the mid-1920s, the fervour with which the extermination programme was pursued had worn off.

Part of the reason for the Indian government's change of mind was the very success it had attained. Bounty hunters and sportsmen, combined with extensive deforestation for cultivation, had wiped the tiger off the map in Sindh and Punjab. Provinces were now left to their own devices. General rewards were abolished in some regions, even as bounties for specific marauders were handed out. Extinction loomed, a distinct issue. It was possible to make out a strong case for protecting the cheetah and the lion, which posed little threat to human life or property. 'For the sake of rarity alone, it seems desirable to protect them,' wrote G. Bower. The very forces that had powered elimination now hesitated. Such sentiments found echoes but had only limited impact on policy. By this time, the lion was extinct in British India and only survived in the princely states of Saurashtra. The cheetah was already in serious decline, and perhaps found no refuge because it was not a key trophy animal. No reprieve awaited the wolf or the leopard, which were seen simply as pests. Conservation did not easily extend itself to the protection of carnivores. Rarity was a precondition for protection but brought no guarantees in the struggle for survival.[5]

It is difficult to determine whether at all bounty killing played a major role in exterminating key large mammals. There is no question that it tipped populations over the edge in places where habitat was under increased pressure. In fact, the numbers of animals killed for rewards were often a good index of the land deforested for agricultural expansion. Over 80,000 tigers, more than 150,000 leopards and

200,000 wolves were slaughtered in the fifty years from 1875 to 1925. It is possible this was only a fraction of the numbers actually slain for rewards; officials recorded *only* those cases where they paid out a sum of money. A delay in claiming the bounty meant the killing went unrewarded and unrecorded. Even if one assumes that one in three such kills were reported in time and the evidence produced, it would be a pointer to very high populations of carnivores in the wild. It would also indicate an astoundingly large prey base, for a tiger may need to kill fifty to eighty heads of herbivores in a year. The actual impact on tigers was limited in parts of central India, the United Provinces and Bengal with their extensive forests and abundant wild prey. Even here, its range contracted to hill forests, marshes and less accessible regions, but extinction did not seem a possibility.

Elsewhere, the pressures may have played a crucial role, especially if the numbers killed off exceeded those born and raised in fresh litters. Here, the extra incentives given to kill females and cubs may have had catastrophic consequences. With the larger carnivores, the reduction of prey numbers may have had an equally powerful role to play in speeding up decline. Bounties were not the sole or the major cause of decline, but they are too important a factor to be left out in a story of the past. In general, the large mammals of the plains and open scrub forest suffered more, with direct destruction being the forerunner of agrarian expansion. In mature tree forest, it was still possible, as long as the cover was intact and prey plentiful, for animals like the tiger to recover from inroads made by hunters. Though remarkably fast breeders, predators like the tiger, which were once a familiar part of landscape, vanished from vast swathes of rural India. As late as 1929, a Bombay-based hunter could shoot dead a tiger that had swum on to the island across the Thana creek. This was the seventh successful tiger shoot on Bombay island in three decades. It was also the last: no more would the striped cats roam this stretch of coastal forest.[6]

Europe had experienced centuries of state-sponsored carnivore killing but it was a new experience in India. If it worked, it was only due to a degree of cooperation by different groups of Indians. Trappers and snarers knew the behaviour and habits of quarry as elusive as the wolf or the tiger. Grain-growing farmers and herders of sheep, goats and cattle had reason enough to take advantage of the system of bounties to settle scores with animals with which they were already in conflict. The limits to power were still much in evidence. In many areas where the confrontation between people and predators was not so intense, extermination did not work. Religious and cultural factors meshed with self-interest. The great change was in the dry and plains country, whose large vertebrates were more susceptible to the new pressures. Perhaps the tensions between lions and livestock were sharper than their equivalent in tiger country. In turn, smaller carnivores like the panther could eke out a living in areas where, with the forest thinned out, the tiger found it impossible to survive. The wider change was a very significant one. Over vast stretches of terrain, the cheetah's sprint and the roar of the lion became a thing of the past. If the tiger did survive, it was also because much of its home was in the hill forests: protected by the British interest in timber. Elsewhere prime tiger habitat like the tarai grasslands of north India or the mangroves of the delta of the river Ganga was inhospitable for humans. The animal was no longer safe where humans lived. A new page had been turned in the story of wildlife–human encounters in South Asia. The Raj left its deep mark on the natural world. Bounty hunting was not the only force at work: killing for trophies reached unprecedented levels, and forms the next part of the story.

Princely Hunts and Royal Preserves

The Big Game Diary of Sadul Singh, Maharajkumar of Bikaner, privately printed in 1936, catalogued his bags over a quarter of a century. In this time, he had ranged far beyond the confines of his desert kingdom in western Rajasthan to shoot tigers in the forested hills of central India, lions in the dry teak jungle of Saurashtra, leopards in Bharatpur and wild buffalo in the Nepal tarai. The rarer a creature, the greater the sense of exultation of the big game hunter. Thus it was that on a March morning in 1920, Sadul Singh's father grew taut with excitement when trackers brought news of a male wild buffalo that they had seen mingling with a tame herd. The party of hunters had travelled all the way to Babia-Bankuwala in Nepal to get a fine head of the great arna, a creature already so rare across north India that the British had restricted its killing for sport. As they waited in a macchan, a platform on a tree at the edge of a plain, Sadul Singh recalled how, 'Father got very excited as it was his first experience of this kind, even though he is an experienced sportsman. He said his heart was thumping as in his early sporting days.' Five years later, the same hunters gave the *coup de grace* to a group of three cheetahs in Rewa, a princely kingdom in central India. The cheetahs, shot from a motor vehicle, were so rare that their shooting was described as 'a great piece of luck'.

The *Diary* is a priceless document because it totals all that Sadul Singh shot over a quarter century. Nearly 50,000 head of animals and a further 46,000 game birds fell to his gun. Among these were 33 tigers, 30 Great Indian Bustards, over 21,000 sand grouse and a lone Asiatic lion. To cap it all, over a thousand of the game animals had been bagged outside India. The Cape buffalo and the black rhino were among the 33 varieties of herbivores of the savannah and jungle of Africa that ended up as trophies in the Bikaner palace.[7]

In the annals of the hunt, the Indian princes stand out in a league of their own. Despite recent attempts to rewrite their role as precursors of modern-day conservation, their record was not always an edifying one. Having submitted to the British as the paramount power in the land, they were treaty-bound to eschew war. Disallowed from taking up arms on the battlefield except in service of the Raj, they used their time and the labour of their subjects against the wild animals and birds of forest, marsh and savannah. From around a century ago, the hunting grounds of the princes acquired new importance. As certain large game animals became more scarce in British-ruled territories, officials from the Viceroy down to the district officer vied with each other for an invitation to the sportsman's paradise that lay in princely India. A brief trip could bring the visitor a rich haul of trophies, to be prepared by a professional taxidermist. If it was a large specimen it could be reported in the record books of big game, preferably by Rowland Ward of London. After the Rebellion of 1857–8, when direct rule by the Crown replaced that of the East India Company, the princes were seen as pillars of the imperial power. The hunt reinforced and symbolized their loyalties even as it enabled them to mingle with the high officials of the Raj. In a racially polarized empire, they were seen as 'honorary whites', whose loyalty, personal bravery and marksmanship was contrasted with the effete urban middle class which was already asking too many uncomfortable questions about the legacy of British rule. It was in what was labelled as 'Indian India' that the hunt was refined and developed into elaborate ceremony. A Viceroy would be pleased to 'get' a tiger or two in a morning's hunt. A bag of five in a central Indian hunting preserve in the 1920s was enough to earn the ruler a number of favours. The famous cricketer Douglas Jardine (of Bodyline fame) went out in the hunting estates of the Raja of Singahi for his big game shoot. The raja could get his guests forty head of swamp deer stags on a single day. The marsh 'the deer inhabited was strictly protected to serve up such large bags when required. It was this that made the rulers classify some key species as game, a resource so critical to their political machinations that it simply had to be used judiciously.

If only a few were allowed to shoot and kill, it did not mean that theirs was an ethic of nature protection. In the third of India that was under the princes, killing was a rite of passage into adulthood, especially manhood, for a number of dynasties. Such rituals still persist among landed elites in parts of central India, though they are often carried out stealthily. Each state or region had its own distinctive styles of hunting and coursing game. Gayatri Devi was a princess of Cooch Behar, a state in north Bengal with perhaps an unrivalled record of big game shoots in all of eastern India. She went out on a shoot at the age of five and 'got' a panther all her own when only twelve years old. When she bagged her first panther, she got a congratulatory telegram from Jai, her future husband. She would later describe it as 'almost as thrilling as the kill'. After she married the ruler of Jaipur, Gayatri Devi would go out on tiger shoots in

Sawai Madhopur (in what is now the Ranthambhore tiger reserve) and on buck hunts with captive cheetahs. The underlings and vassals of the princes also aped their habits and to the extent possible, their lifestyles. Ruling clans and castes were zealous about the exercise of their hunting privileges. But their ethic was one of conquest and use, with nature as a surrogate for the political power they had lost.[8]

In fact, in their zeal for large bags, many princes outstripped their British masters, and it would be anachronistic to see them as modern-day conservationists. Some did not even spare tigresses, despite pledging to abide by hunting ethics. Before 1900, George Yule with 400 tigers and Montagu Gerrard with 227 had among the largest British bags. They were to be outstripped by men like the ruler of Udaipur and the Raja of Gauripur with 500 each. The Nawab of Tonk shot a total of 600 tigers. He stopped killing tigresses after he got to the score of 150, but did not prevent his subordinates from shooting them. Even in the 1940s, his chief hunter would gladly report how he had 'included a tiger shoot' in the itinerary of a marriage programme. Ramanuj Saran Singh Deo of Sarguja was to hold the all-time record of over 1100 tigers in his lifetime. By the time he stopped shooting in the 1950s, he had another less known record against his name: of over 2000 panthers or leopards. He had even killed thirteen of them in one night, luring them to specially built platforms where goat kids and mongrels were tethered. Eager for a meal, the big cats came to their deaths.

* * *

At times, the shikar or hunting department could use brute force to give teeth to the law. Colonel Kesri Singh of Jaipur boasted that he had participated in a thousand tiger shoots. Sometimes, the Colonel turned to other game. When he found tribals poaching in an antelope preserve, he placed iron hooks on select trails to injure the soles of their feet. His forest guards were asked to find 'a nameless grave among the dense jungles' for those who disobeyed. The rulers could race their cheetahs or shoot buck in the reserves, but these very areas were closed to their subjects for hunting if not for grazing. Worse off were tribals and caste Hindus who had to take part in the great drives or *hakkas* for large game. A tiger or a panther when chased can turn deadly. And in most princely hunts the big game hunter sat safe on a raised platform or on elephant back. It was the beaters who were most at risk. A sixteen-year-old boy mauled by a panther in the Gir Forest or a man succumbing to injuries inflicted by a cornered tiger in a central Indian forest: these are the kinds of notations in an old hunting diary. No wonder forcible labour dues or *begar* and curbs on access to forests became an explosive political issue in many princely states by the time of Independence in 1947.[9]

Many princes initiated protection to secure good trophies. In the early phase, tigers or other royal game were guarded jealously to assert the rights of the local ruler. As bag sizes and record specimens began to matter, more refined techniques of management came into play. Sarguja's hunting grounds were divided into different zones. Some tracts were out of bounds for grazing cattle, while other forests not maintained for tigers were open to livestock. Some princes observed bag limits to perpetuate the existence of the hunt: wild animals were being transformed into 'game'. Others did not give out bounties for carnivores as vermin. Yet such restraints obviously had limits. Sport hunts on a large scale denuded many princely states of their fauna. Modern weapons were pressed into service, each being vetted by the British Resident or Political Agent, to ensure they were only being used against the denizens of the woods, not to stir up trouble. If much fauna and flora survived, it was not due to princely prudence in the hunt, but because the richest plains regions had been lost to the British.

* * *

Princely India left behind a mixed legacy. There is no doubt that many of today's famous nature reserves from Gir in the west to Bandipur in the south had their origins as royal hunting reserves. But to see the princes' efforts as conservationists in present-day terms would go against their own record of their deeds. Many exceeded the

British in the lust for trophies. They used the rules of the hunt to oppress their subjects, at times endangering the latter's lives and much more often, offending their sense of human dignity. The political assertion of their subjects coincided with a deepening lust for trophies on the part of the rulers. In the twentieth century, it was Indian princes who held the record bags for the numbers of game shot dead. Several of them outstripped the scores of their British masters. No wonder that this pyramid built on such a narrow base came under threat once imperial power retreated from South Asia. The heyday of the princely hunt was over, but it still had some life left in it, as subsequent events would show.

Notes and References

1. National Archives of India (hereafter NAI), Delhi, Home (Public), January 1875, A, nos 286–311, no. 297: Mjr. Tweedie, Hyderabad to Secretary, Home, 5 January 1874.
2. K. Sivaramakrishnan, *Modern Forests: State-making and Environmental Change in Colonial Eastern India*, Delhi: Oxford University Press, 1999, pp. 91–100; T.T. Cooper, *The Mishmi Hills*, London, 1873. Delhi: Minal Publications, 1995, p. 63.
3. J. Fayrer, *Destruction of Life by Wild Animals and Poisonous Snakes in India*, London: The Royal Society of Arts, 1878.
4. G.P. Sanderson, *Thirteen Years Among the Wild Beasts of India*, London, 1878, Delhi: Minal Publications, 1983, pp. 267–8; M. Rangarajan, 'The Raj and the Natural World: The War Against "Dangerous Beasts" in Colonial India', *Studies in History,* Vol. 14 (1998), pp. 265–300; on wolves, see F.J. Richards, *Madras District Gazetteers, Salem*, Madras: Govt. Press, 1914, p. 37; Fayrer, *Destruction of Life*, p. 11.
5. Anon., 'Big Game in India', *The Times*, London, 14 December 1907; NAI, Home (Public), March 1908, A, nos 27–53: Collector, Shahpur to Chief Sec., 21 July 1904.
6. Rangarajan, 'The Raj and the Natural World'; S.H. Prater, 'On the Occurrence of Tigers on the Islands of Bombay and Salsette', *Journal of the Bombay Natural History Society (JBNHS)*, Vol. 33 (1929), pp. 973–4.

7. Sadul Singh, *The Big Game Diary of Sadul Singh, Maharajkumar of Bikaner, 1910–36*, Privately printed, Appendices and pp. 15–16 and 59.
8. Arjan Singh, *Tiger! Tiger!*, London: Jonathan Cape, 1984, pp. 48–9; Gayatri Devi, *A Princess Remembers: Memoirs of the Maharani of Jaipur*, Delhi: Jaico, 1973, pp. 99–100.
9. On Wankaner, see Digvijaysinh, *The Eco Vote: People's Representatives and Global Environmental Problems,* Bombay: Prentice Hall, 1985, pp. 63–4. Sadul Singh, *Diary*, pp. 34–5, 125–6; Kesri Singh, *Hints on Tiger Shooting*, Bombay: Jaico, 1969, pp. 114–15.

C.E.M. RUSSELL

THE TIGER (*FELIS TIGRIS*)

Despite the facts that so many English people have relations and friends earning their living in India, and that so many Englishmen of means now visit that country, it is surprising to find how great is the ignorance which prevails at home regarding the big striped cat who is the subject of this chapter.

English people are wont to believe that tigers are common in India, and that a man has only to be keen on shooting, and to desire interviews with these interesting felines, in order to obtain plenty of skins.

As a matter of fact, however, the truth is (alas!) exactly the reverse, and every sportsman has ascertained the falsity of the pleasing fiction so soon after his arrival in India as his circumstances may have rendered it possible for him to go out tiger shooting. Many keen sportsmen have been out in India for a number of years, have spent a good deal of both time and money in trying to bag tigers, but have not succeeded in slaying even one.

The fact is that tigers are necessarily rare animals, for they prey ordinarily upon other *fera naturæ*, and it follows from this that were they to become plentiful in any one locality, the game would be killed off and the tigers forced to migrate. The ultimate result would undoubtedly be that the tiger would become extinct. Nature, however, maintains so even a balance that this danger has been completely guarded against; and, although the eventual extinction of the tiger is probable, there are so many vast solitudes but rarely inhabited by man in the immense continent of India that, although he is an uncommon beast, his extermination is still very far off.

But for the havoc wrought by man amongst the wild animals upon which the tiger preys, there would no doubt be food for more of the latter; but the fact being that the country bristles with guns in the hands of natives who shoot only for the pot, and who spare neither females nor young, and as moreover there are so many meat-eating castes that shooting venison for sale is a profitable business, deer, etc., will soon be exterminated in forested areas near villages; and the tiger, his food supply being cut off, will be forced to seek haunts more remote from the borders of civilization, where game may still exist.

Unfortunately, a reward, which is in Mysore as high as fifty rupees, is paid for the destruction of each tiger. Now when we reflect that a forest guard in Mysore draws pay at the rate of only six rupees per mensem in most localities, we can well imagine how profitable a business it must be for a man of his class to shoot deer, etc., for the purpose of sale, and to occasionally shoot a tiger for the sake of the reward.

This, then, is another reason why tigers are even rarer than Nature requires them to be, for, owing to the scarcity of the tiger's natural food which is fast being exterminated by native gunners, the former are compelled to take toll of the villager's cattle, and then comes the chance of the native, who, lying *perdu* in perfect safety in a tree, watches for the return of the slayer to feed upon his victim. Should the tiger so return, he is either killed, wounded, or missed, and seriously scared by the would-be bagger of so many rupees! I have, however, heard of a case in which the ambushed native was so struck by the imposing appearance of the animal, to shoot which he was watching, that he was too scared to fire at all, and the tiger ate the carcass before the eyes of the man, who remained all night in the tree, afraid to descend!

If Government were to abolish the reward, natives would no longer have any interest in shooting tigers, except, of course, any such as might become great oppressors of any one village, in which latter event the beast would get very short shrift. In my opinion the time has come when the reward *ought to be abolished*, for, while tigers are so rare, guns are so very common, that there is no fear of any community, which might suffer heavily from the rapacity of a tiger, failing to take steps to rid itself of him.

* * *

TIGER SHOOTING IN SOUTHERN INDIA AND HINTS TO BEGINNERS

Everyone fond of big game shooting is very keen to bag a tiger whenever the opportunity may offer, and the rarity of the animal only enhances the sportsman's anxiety to succeed in each attempt.

As a matter of fact, however, considered as a form of *sport*, tiger shooting cannot be compared with bison and elephant shooting, or with sambur and ibex stalking on the hills. The reason for this is that the sportsman's own part in it is so very small a one, by reason of the number of accessories—it may be elephants as in Bengal, or beaters as in Southern India—which are required, and without which, unless he should happen—a very rare piece of good fortune indeed—to meet with one accidentally when stalking in the jungle or on the hills, or to successfully sit over a kill, he has no chance whatever of bagging a tiger.

Of howdah-shooting from elephants, as practised in the expanses of reed and high grass in Bengal, Nepaul, and Assam, I have had no personal experience, though my father (who was in the Bengal Civil Service, and had great opportunities for the sport) did much tiger shooting by this method.

In Southern India, the sportsman is usually posted on a rock, tree, or shooting ladder, and a crowd of natives—some of them employing horns and tom-toms (native drums)—endeavour to beat the tiger up to him.

This method is often, somewhat erroneously, termed 'tiger shooting on foot,' though, if the tiger should go on wounded after the shot, he must be followed up on foot; and this operation is the most dangerous one which the Indian sportsman is ever called upon to perform.

Another method by which a tiger may be shot is by watching for his return to feed upon the carcass of a buffalo or a cow which he has killed; and, unless it be adopted under certain circumstances, *e.g.*, when a tiger has killed in a large tract of forest in which beating would be out of the question, a chance (a poor one though it be) is sacrificed.

Tiger Shooting with Beaters

Wherever the jungles are not too large and continuous, this method is the one which is most frequently successful. A great deal depends upon the cover in which the tiger is supposed to be lying up after a heavy meal of beef. If this be of considerable extent, and especially if intersected with ravines, some of which diverge laterally from the main longitudinal nullah, in the absence of men well accustomed to the work, and of a large contingent to act as stops, the odds against bagging the beast are heavy.

If, on the other hand, there should be but one ravine, or a stream of water flowing through the cover, and the latter be of reasonable dimensions, the chance is a good one.

The first thing that a tiger does after eating a heavy meal is to make for the nearest water, to walk right into it, and to drink deeply. He then, unless he should feel inclined for a second feed, betakes himself to the nearest suitable cover where he can obtain cool shade, and from which water is not far distant.

He has generally eaten both hind-quarters of his victim during the first night, and he intends, after sleeping off the effects of his heavy gorge, to return to the kill, and to devour the remainder of the flesh on the succeeding night.

Bearing the above points in view, and with the remark that the hot weather, *i.e.*, from February to May, is *the* best time for the sport, we will now discuss the *modus operandi* of, say, three or four guns, who may have decided to form a party to shoot tigers in any given district.

It is essential that three or four natives belonging to the district, who are keen upon securing success (or, at least, upon earning rupees as a reward in the event of good sport) should be engaged as shikarries. These men must know the country and the people thoroughly well, be active and willing, and also ready to carry out all orders *promptly, and to the letter.*

It is further essential that unless one at least of the party be a Government official belonging to the district to be worked, the sportsmen should invoke the assistance of the authorities by writing a polite note to the Collector (or calling upon him, should that be practicable), and asking him to kindly issue orders to his subordinates for their assistance. (In Mysore, and in non-regulation provinces in India, the head of each district is called, not 'Collector,' but 'Deputy-Commissioner.')

Without the assistance of the authorities, it is in many places well-nigh impossible to induce villagers to turn out to beat, and in fact in too many localities, owing, in the first place, to the extreme general leniency of Government towards the natives, and in the second, to a too often rabid and scurrilous native press (recently, however, somewhat brought under the curb), the natives appear to take the keenest delight in thwarting and obstructing an Englishman in every possible respect. The party must, therefore, in the first instance, and in ample time, invoke the assistance of the authorities, and should their request for the same be met with even a churlish and half-hearted acquiescence, they had better decide to leave that locality alone and to try another. In many districts it is necessary to obtain a licence from the Collector to shoot in forest reserves, and, during the hot and dry weather, this is often refused in the interests of forest fire-protection—*verb. sap. sat.*

Supposing, however, that all has gone well, and that the Collector, or Deputy-Commissioner, as the case may be, has issued the necessary orders to his subordinates, the next matter to be settled is the plan of campaign.

It is at this stage, and not until now, that the local native shikarries before alluded to should be engaged, and in consultation with them the sportsmen will decide upon the best locality for their first camp.

It is presumed that each member of the party has brought at least one horse or pony, and that the one who is in charge of their commissariat has provided all camp requisites, as well as a sufficient supply of provisions, liquor, and soda-water, to last them for the trip; or that it has been arranged that consignments of the three latter shall meet them from time to time at pre-arranged places.

The spot to be selected for the first camp should, if possible, be a central one, with jungles frequented by tigers within easy reach on all sides, and it must be close to good water, and sheltered from high winds.

* * *

INCIDENTS IN TIGER SHOOTING

In the autumn of 1881 I left Assam and went to Mysore, where I had been offered, and had accepted, an appointment in the Forest Department.

I had previously met the author[1] of *Thirteen Years amongst the Wild Beasts of India*, quite accidentally in Calcutta, and first heard of his book (which, of course, I at once purchased) from himself, and I was charmed at the prospect of going to a country where sport is obtainable on foot; whereas in Assam, without employing tame elephants, and consequently incurring much expense, a sportsman can do nothing.

In 1882 I had many opportunities of big game shooting, and I bagged my first elephant and some bison, deer, and pig, but did not even *see* a tiger.

On the December 14, 1883, on my return from inspection duty to my camp at Naganipur, news was brought me that a tiger had killed a bullock at no great distance. I hurried off to the spot, and sat on the ground on one bank of a shallow nullah in which the carcass lay, but up till dusk, when I returned to camp, the tiger did not appear.

The next day I went to see if he had visited his kill during the night, and found that he had done so, and had moreover dragged the bullock to some distance, leaving it in a very dense, thorny thicket.

I had a mechan put up in a tree near, and caused the carcass to be dragged from under the dense canopy of thorn, and left in the open in front of my tree.

During my vigil, a jackal came and loafed round the kill in an aimless sort of way, and at some distance from it, as if he had not seen it at all, and then disappeared in the jungle.

Presently, having obviously made a complete, or almost complete circuit, he reappeared from the direction in which he had first shown himself, walked up to within a few paces of the defunct bullock, and then jumped backwards, as if alarmed. He repeated this performance several times, going a little nearer to the coveted beef each time, and then craned out his neck as far as he could, and gradually and cautiously touched it. Directly he touched the kill, all his fears appeared to evaporate, as he evidently made up his mind that had the tiger been anywhere near, his preliminary acrobatic performances would have elicited at least a warning growl. He thereupon set to work in a very business-like way, and tore the stomach open, when a most fearful stench rose in the air and seriously incommoded me. I squirmed slightly on my mechan, the jackal gave one upward glance, bolted, and I saw him no more.

At a little before six o'clock, while it was still quite light, I saw the tiger advancing slowly through the thicket in which the kill had been placed, and from which it had been dragged a few paces by my orders, so as to render the way clear for a shot should he come. He looked backwards only once, and then came right up to the kill. I was afraid of his seeing me and dashing off alarmed if I raised myself before his head was hidden by my mechan, but as soon as it was out of sight, I elevated myself and my rifle and fired down upon him. As the smoke cleared away, I saw him slowly disappearing, as if he were dragging himself along with difficulty, and I fired a snapshot, which apparently missed. I got down as soon as my men, hearing my shots, came with a ladder, and then I found that my first bullet had grazed a green stem on its way to the tiger, who, however, had evidently gone off severely wounded.

I returned to camp, and wrote at once to Government requesting three days' casual leave, during which I hoped to bring the wounded beast to bag.

I had two dogs in camp with me, one of which, 'Carlo' by name, was a nondescript animal, regarding whose origin, and the number of breeds contributing to whose composition, it would have puzzled the doggiest man alive to form even the faintest opinion. He was formerly the property of an Ootacamund native shikarrie, and had been much used in sport on the Nilgiri hills. He was kindly procured for, and presented to me by Mr (now Colonel) N.C., the hero of the boxing match with the wounded bull bison which is elsewhere related. 'Carlo' was a capital dog out shooting, in spite of his having lost an eye before he came into my possession—whether by the horn of a sambur stag, or by the quill of a porcupine, I never learnt. My other dog, or rather bitch, was a novice who rejoiced in the name of 'Puppy', and she too was a mongrel, with a predominating touch of fox terrier in her.

Next morning, accompanied by a good many men and by my two dogs, I set out to follow up the wounded tiger. We proceeded to the spot where he had been wounded, and followed up the blood trail, which led through terribly thick stuff, in which the danger was extreme, the advance being of course proportionately slow and cautious. We had in some places to even cut our way. Presently, we heard the tiger groaning in front, but could not see him. The tracks entered a lightly jungled ravine which debouched into a stream, the latter in its sinuous course permeating many portions of the Naganipur jungles. Telling the men to wait till I had got on ahead, and then to throw in stones and to loose the dogs, I went down to the spot where the ravine met the stream, and then I saw by the tracks that the tiger had already crossed, so we had to follow up again. After some distance they led into a very densely jungled, but narrow nullah, and I directed the men to let me get well ahead, and then to come along it on both banks, throwing in stones, and keeping the dogs at work, but on no account to themselves enter the ravine.

I accordingly went ahead with a man carrying an 8-bore ball gun, while I took my .500 express, and making a detour, we struck the nullah bank some distance down, when, taking the precaution to relieve my attendant of the spare gun, I placed the latter resting against a tree. I stood on the bank and waited the issue of events.

The beat began, and by-and-by I heard old Carlo barking, and very shortly afterwards, out came the big, round head of the tiger, on my side of the nullah, and only some twenty or thirty yards off. His head alone was visible, but he apparently wished to break out at the side, in which case he would have given me a broadside shot; when, as bad luck would have it, my attendant, overcome with fear, fell, and the tiger, his attention thus drawn to my direction, instantly spotted me, and with a 'woof!' he started forward at me. I fired immediately, and he disappeared in the nullah. I at once shouted to the men to retreat, and then proceeded cautiously to the spot at which he had vanished.

There I saw, in the erebean darkness caused by the dense shade, two fiery balls at the bottom of the deep ravine. I made sure these were the eyes of the tiger, and, aiming carefully between them, I fired, and then found that what I had thus mistaken for eyes, were but two gleams of sunlight which had penetrated the blackness of the gloom below, and that the tiger had gone back down the nullah. We followed, and found that he was in anything but an amiable temper, as he had *en route* picked up a thick piece of creeper stem and had bitten it, leaving blood upon it. We carried the tracks back across the stream until they entered a very dense thicket, and there I pegged a piece of paper to the ground to enable us to find the exact spot on the morrow, and then returned to camp. Heavy rain came on, and I almost despaired of our ability to distinguish the tracks next day as the rain would certainly wash away all bloodstains from the trail.

Next morning we went to the spot at which we had left the tracks, and the men began cutting the jungle to enable us to get through, when, from close in front of us, we heard a heavy animal moving off. Taking the men with me, I made a détour, and we found, in front of the thicket whence the sound had proceeded, a small piece of perfectly open ground, in advance of which lay a dense patch of sigee thorn which came nearly down to the ground, and so allowed of no view.

As we approached this second thicket, a deep growl sounded from under it; I told the men to stand firm, and they behaved well.

In front of the impenetrable cover flowed the stream, and I put the men up trees in a semi-circle, the extremities of which touched its banks, and directed them to give me time to cross its bed and to ascend the further bank, after which they were to shout, and fire shots from a shotgun which I had placed in the hands of one of their number.

I crossed and took up a position on the further bank, and the shouts and shots rang out without any effect; and we then found, on examining the thicket, that the tiger, after growling at us, had crossed the stream and gone on, and that he was therefore not in the beat at all when our arrangements were completed. Under the thorny canopy, we found several blood-stained forms where he had lain during the previous night; and he must have moved from this thicket to the one from which we heard him moving off (and at which we had left his tracks on the previous afternoon), after lying for a long time—probably all night—in the former. Evidently he was desperately wounded. We followed his trail for some distance after this, and found that he had crossed a small hill, during his progress over which he had been obliged to lie down several times.

That the tiger could not have got away, had I a steady elephant, was certain, but that we had no chance of bagging him, in the dense thickets in which he always took shelter after crossing a bit of fairly open jungle, was equally sure; and that to press him at such a disadvantage would lead to a fatal accident, was most probable; so that at last we decided to return to camp, and to send out thence to the neighbouring villages to procure all the nets which we might be able to obtain, with the aid of which we hoped to bring him to bag on the following day. I therefore went back to my tent, and that evening I sat on the ground in the jungle, with a kid picketed in front of me, and bagged a panther.

Next morning, having succeeded in obtaining only a few nets, we went out and again took up the tracks, which soon led into a large and very dense thicket. Six or eight times the number of nets at my disposal would have been necessary to enclose the same, which was situated on the bank of the stream, where the latter made a bend at nearly a right angle. I therefore put up the nets across part of the base of the enclosed triangle as far as they would go, and from their termination stationed men up trees to the stream on the other extremity of the base, and also along the portion of the bank which was out of my sight, while I crossed the sandy bed and sat on the further bank at the apex of the triangle. Presently, shouting and firing of shots began, and continued with vigour for some time, but nothing appeared. Carlo, however, had gone off to perform a little personal investigation, and I soon heard him barking vigorously under a tall banyan tree which I could see from my post.

Upon this, as shouts and gun-shots were ineffectual to move the tiger, and thinking also that he might be lying dead under the banyan tree, I decided to go and look him up inside the thicket; and so, taking my own position in the centre of a line of men armed with spears, I followed the still distinct blood-trail, the men with their spears beating down the jungle as we advanced. After a time, from almost under the spears, up jumped the tiger, who went off with a loud 'woof!' Not one of us saw him in the dense cover; but the spear-men retired as if but one man! After ineffectual attempts till evening to obtain a sight of the tiger, I had to give him up and to return to camp.

On the following day—the last of my leave—I went again to the thicket, but the animal had left it, and we were unable to trace him, so I was obliged to abandon the wounded beast—very much to my chagrin.

Editor's Note

1. *The author mentioned here is G.P. Sanderson, a British officer who served in Mysore during the mid-nineteenth century.*

K. Ullas Karanth

F.W. CHAMPION

PHOTOGRAPHING TIGERS

There are numerous ways in which one can attempt to produce photographs of tigers in the wild, and I will now discuss these in detail in this chapter. The most obvious method of all is to locate a tiger in a definite piece of jungle and then to drive him out by means of elephants, or a line of men, flanked by 'stops', past a fixed point at which the photographer is located. This is the ordinary method of shooting tigers in India, but it has little to recommend it to the animal photographer. Firstly, under such circumstances, the tiger is fleeing for his life from his dread enemies, so that any photographs one might produce as a result of such a beat would not represent a tiger under natural conditions, where he is the pursuer and not the pursued. This objection alone rules out this method at once for anyone who wishes to picture tigers living their own lives as kings of the jungle, and hence I have spent little time in the pursuit of such photographs. Apart from this, however, there are other very serious objections to the employment of beating for the production of photographs of tigers. In the first place, even if the beat be extremely carefully and quietly carried out, the tiger will most probably come out at a fast walk, and he may rush out at a gallop, particularly when he has to pass the open spaces which he hates, but which are essential to the photographer. An exposure of at least 1/150th of a second is necessary to stop movement in a tiger walking, whereas it may run up to 1/1000th of a second to produce a satisfactory picture if he be going fast, and, except in very rare cases, such short exposures are absolutely impossible in the bad light of the Indian jungles. Again, no tiger will cross the open in a beat if he can possibly sneak out under cover, so that, even if he comes out sufficiently slowly to make an exposure possible as regards movement, he is almost certain to come through shady grass or bushes, in which the exposure required for the production of a fully-exposed negative may run into seconds. And who can hold a reflex camera still for seconds, even if the tiger himself would remain perfectly motionless for such a period of time? In Nepal, tigers are sometimes shot by ringing with elephants, and, given a sufficient number of elephants, it would be perfectly possible to ring them in open country suitable for photography. Indeed, daylight-photographs of tigers have actually been produced in this way in Nepal, but, as the tiger might just as well be inside the bars of a cage as inside an almost impenetrable wall of elephants, such photographs make no appeal to me.

Beating having been ruled out, the next most obvious method of photographing tigers is to obtain a kill and then to sit quietly in a machan in the hope that the tiger will arrive in bright daylight and thus enable an exposure to be made. My personal experience is that tigers in the heavily-shot-over jungles of the United Provinces generally arrive well after dark, although some may come at dusk and a very few at any hour of the day. I have sat in a machan with and without a camera many times, but I have never yet had a tiger come in a light good enough for a picture to be produced. It is to be remembered that the light in the Indian jungles, even under the best conditions, is such that a reflex exposure can be made only in direct sunlight and many kills are located in such dense cover that, even if the tiger were to arrive at noon, the production of a satisfactory picture would still remain almost impossible. Also, if one sits in a machan, one must sit behind a screen of leaves and these leaves and intervening branches are anathema to the photographer even though they may be little hindrance to the sportsman. I have several times been told by sportsmen, who were not experienced animal photographers, that they have had tigers come in a beautiful light, in which they could easily have taken fine photographs—if only they had had a camera with them! I wonder! None but those who have actual experience of photographing wild animals in dense forest can realize how many apparently insignificant little details are essential to the making of satisfactory photographs, and I very much doubt if many good pictures of tigers will ever be produced in this way, although I admit that a perfect opportunity may possibly present itself once or twice in a lifetime. As a regular method of attempting to photograph tigers I consider this also to be of little practical value.

A third method is to move quietly about the jungle, on a tame elephant, in the morning and evening in the hope of a chance meeting with a tiger. This is called 'ghumming' in India and is of value, particularly in the morning, evening light being very weak photographically, and should bring success if persisted in for sufficiently long in tiger country. I have not obtained any of my tiger-pictures in this way, although it has proved exceedingly successful with some other animals; but, if all alarm-cries are immediately followed up, it is undoubtedly a possible method of photographing tigers and one which should have brought me more success than it has done. On one occasion, in the hot weather, I came across a tiger, lying half-asleep in the sun at about 10 a.m., and completely blocking the path along which I was proceeding at the time. We first saw him at a distance of about 30 yards, and, as his head was turned in the opposite direction, it was obviously an excellent opportunity to stalk right up to him. My mahawat on that occasion, however, a sullen and bad tempered man, argued that I must shoot him straight away, and, when I refused, he retaliated by taking the elephant up fast and noisily, thereby causing the tiger to bolt at once. This is the great difficulty which always has to be faced if one wishes to use tame elephants for the purpose of big-game photography. The mahawats have been trained from boyhood to expect the sahib eagerly to seize every opportunity of shooting animals like tigers, and they simply cannot understand anyone letting such good opportunities slip for the sake of a paltry photograph, which, to them, means nothing whatever. A mahawat loves to boast among his fellows of the number of tigers that has been shot from his elephant, and he is always thinking of the 'bucksheesh' which accompanies the death of a tiger. The photographer can overcome this later difficulty by giving large tips whenever he obtains a successful photograph, but, even then, nine mahawats out of ten will secretly despise him and think him quite unworthy to sit on their elephants.

There is a much better method of photographing tigers, nevertheless, than those already mentioned and that is to obtain a tiger-kill in country suitable for photography and then to attempt to stalk the tiger, on a tame elephant, when he is resting in the heat of the day. This method requires an intimate knowledge of the country and drinking-places, but it is one which is sure to bring success, sooner or later, if pursued with sufficient care and forethought, and always provided that one has a perfectly staunch tame elephant at one's disposal. The normal habit of tigers, after eating, is to drink at the nearest pool, and then, provided they are undisturbed and the country is suitable, to lie up for the day, either near the kill or near the drinking-place. It thus helps in locating the tiger to arrange for a kill with water and a suitable lying-up place near by, and this can be done by a judicious choice of places for tying out buffalo-baits. Needless to say, it is of vital importance that the men who go to bring in the baits in the morning should make no noise whatever and on no account should follow up the drag, in the event of any of the baits having been killed. Also, it must always be remembered that a tiger's choice of a retreat for the day will depend very largely on the season of the year: in the summer he will almost invariably seek out the densest cover, with water in the immediate vicinity, whereas, in the winter—at least in so far as the extremely cold Himalayan foot-hills are concerned—he will often lie right out in the sun on some open hill-side, from which position he can command a view of the whole country, including the approaches to his kill. It thus follows that neither of these places is of much use photographically, the former being much too dark and shady and the latter so open that a close approach becomes very difficult. The ideal is a compromise between the two: open tree-forest, which permits the entry of sufficient light for instantaneous photography and also, at the same time, enables one's tame elephant to approach slowly and partly under cover, so that the tiger may be deceived into thinking that the approaching elephant is a wild one having no connection whatever with man. I have succeeded in getting right up to tigers a number of times in this and other ways and one or two experiences are described in full in other chapters.

I have now discussed various ways of photographing tigers by daylight, but, after years of continuous effort, I have found that one obtains very few pictures in proportion to the amount of energy expended. In addition, pictures of tigers by daylight are not truly representative of such nocturnal beasts, who are half-asleep most of the day as they rest from the efforts of their previous night's activity. Truly to represent a tiger as the dread terror of the jungle-night it is obviously necessary to photograph him at that time and the only way to do this is to take up flashlight photography, of which many photographers fight shy as being expensive, complicated, and giving poor 'soot-and-whitewash' pictures. Expensive it certainly is and it may involve a greater knowledge of photography than is possessed by the ordinary owner of a kodak; but I think that, with proper apparatus, it is possible to produce, if anything, better pictures of many wild animals by night than one can ever do by day, and this for the obvious reason that one can take photographs by flashlight of animals in motion in a way which is absolutely impossible by daylight in the dense and shady Indian forests.

The pioneer of flashlight photography of animals was the German, Schillings, who, considering the poor apparatus available at the time, produced some magnificent pictures of African animals, 30 or more years ago. I well remember how I used to delight in his pictures when I was a boy, and, since then, there have been others in the field, the foremost being Major Radclyffe Dugmore and the American, Martin Johnson, both of whom have produced fine flashlight pictures of lions and other big-game animals of Africa.

At one time flashlight work was very unsatisfactory, owing to the unreliability of the flash-powder available, and also to the extreme difficulty of synchronizing the exposure of the shutter with the moment of maximum intensity of the igniting flash-powder. Better flash-powder is now available and the latter difficulty has been overcome by the use of electricity, combined with the excellent apparatus perfected by Mr. William Nesbit, of which details are given in the special chapter on photography at the end of the book. This apparatus greatly simplifies flashlight work, and I have to thank Mr Nesbit for having given me the chance to produce the flashlight pictures contained in this book. Without his apparatus I should not have obtained nearly so many pictures, and all keen animal-photographers owe a great debt of gratitude to the inventor of an apparatus, which opens up such great possibilities of fascinating camera-hunting to those who thoroughly master its details, and who will not be deterred by the inevitable failures which must accompany all flashlight work. Indeed, if flashlight photography were too simple, it would soon cease to appeal, since it is one of the hard facts of life that a thing which comes too easily soon tends to lose its fascination. As this apparatus is fully described in another place I will say little about it here. Briefly, the principle is that the flashlight is fired electrically, either by the photographer or by the subject, and the force of the igniting powder acting on a movable rod enables the shutter to be released at the moment of maximum intensity of the light, thereby permitting the use of such extremely short exposures as 1/200th of a second. It is to be noted that the flashlight can be fired either by the photographer or by the subject, and this places a tremendous power in one's hands—a power which one can adapt to the needs of the particular subject which it is proposed to photograph.

Photographers in Africa have apparently found automatic flashlight work, in which the animal itself fires the flashlight, to be of little use for photographing lions, largely owing to the difficulty of other creatures firing off the flashlight before the lion has had a chance to do so; but, despite innumerable failures from one cause and another, I have found it to give excellent results with tigers—provided always that one still carries on, even when a continuous run of disappointments has made one wonder if there is ever going to obtain any results at all.

To the keen student of animals, it is naturally of much greater interest to sit over a tiger-kill in order to enjoy the extreme pleasure of watching the great beast approach and then to fire the flashlight himself at what he considers to be the best moment; but this, the ideal, cannot always be done for a variety of reasons which I will now discuss. Firstly, it is almost impossible to sit quite still all night, and, if the tiger does not come until late, one may get tired and slack, which results in inevitable movements frightening the tiger away from his kill altogether. Secondly, on a dark night, one can see nothing whatever and has to guess by sound whether the tiger is in a good pose or not—or even if it is the tiger at all—in order to decide upon the moment to fire the flashlight. This difficulty can, however, be overcome to a limited extent by switching on an electric torch just before firing the flashlight. Thirdly, as is described in the chapter on tigers and their prey, constant all-night sittings in the unhealthy Indian jungles undermine the constitution and, sooner or later, are certain to wreck one's health with fever. Fourthly, a Forest Officer, like many another, is a busy man, who tries to carry on his hobby at the same time as his work, and the latter is bound to suffer if he spends many nights out in the jungle. Again, pictures like that on plate xiv[1] can be produced only automatically, since a tiger's approach is absolutely silent and I had to put up the flashlight apparatus at least fifty times before I succeeded in getting a tiger in such a position. If I had sat up all night, for fifty nights, in the pursuit of this picture, I should probably not have heard the tiger coming when he did at last arrive and I should certainly have been in my grave long before this book ever saw the light of day. Lastly, and this is an important point, even if one does sit up all night, one still has very little control over the resultant pictures, since the camera must be fixed and focussed previously and if the tiger comes the wrong way unexpectedly, or appears too suddenly, one may yet have the extreme mortification of seeing all one's trouble come to nought. When on leave, or on special occasions, I would thus recommend firing the flashlight oneself on the arrival of the quarry; but, as a general rule, I am of the opinion that automatic work is inevitable and that the amateur who has his daily work to do will obtain very few pictures without its use.

The usual procedure is to obtain a tiger-kill and then to approach very quietly and carefully in the middle of the day, thus making certain that the tiger does not watch the whole proceeding. When I first started this form of photography I used to erect my flashlight apparatus late in the afternoon, in order to minimize the danger of its being fired off by birds and jackals, and I was astonished at the number of times the tigers never returned to their kills again. I have since found that many of these failures were due to the fact that the tiger had seen what I was doing, and I am convinced that many sportsmen, who fail to shoot tigers from machans, fail largely because they erect their machans too late in the day or too noisily. A tiger's period of minimum vitality is from about 11 a.m. to 3 p.m. and one should naturally take advantage of this as far as one can. Having reached the kill without scaring the tiger, and having made absolutely certain that he is not watching near by, the camera should be carefully focussed, the flashlamps erected, the shutter adjusted and tested, and finally a trip-wire arranged so that the tiger will push or pull it unknowingly when approaching or eating his kill. The choice of site and adjustment of this trip-wire are by far the most important points in automatic flashlight work as upon them depend the success or failure of one's efforts. Almost every kill needs to be treated in a different way: sometimes it may be completely surrounded by the trip-wire, sometimes one can be fairly certain of the tiger's probable line of approach, and sometimes the risk from birds, jackals, and other creatures is so great that it is better to erect the apparatus in some other place, where the tiger is likely to pass during the hours he spends near his kill. The picture of a very large tigress, shown on .plate xxviii, is an example of this. This tigress killed three of our baits within a very short time and always left the bait where it was tied, without taking it away into the jungle. These baits were all tied at cross-roads, or at the junction of a *sot* with a road, so that, in every case, they were found by jackals and numerous, other creatures soon after the tigress had left. It was thus quite useless to erect the automatic apparatus in places where it was certain to be fired long before the cunning tigress could be expected to arrive. The third kill was in a broad river-bed, below a high bank, and experience has shown us that tigers generally approach kills by hugging the foot of such banks. For this reason we arranged the apparatus about fifty yards away from the kill, and put the trip-wire across a small path, after having slightly blocked other paths, with the result that we obtained this picture of the tigress approaching a kill, which had already been visited by numerous other creatures. Her expression suggests her disapproval of these other visitors and vividly shows how automatic flashlight photography can be employed in spite of the constant danger of other animals spoiling the chance. Having decided how and where to place one's trip-wire, one carefully camouflages the whole apparatus with branches and leaves, and tries, as far as possible, by judicious arrangements of thorns and branches, to guide the quarry into the wire, being very careful not to over-do such blocking or the tiger may desert the kill entirely. Many, many times will one fail, but one can and does succeed at intervals, and the few prizes one secures more than compensate for the oft-repeated and at times heart-breaking series of failures, which are inevitable in a form of big-game photography which many have abandoned as useless.

I have heard it suggested that photographing tigers by automatic flashlight is a very poor form of sport, comparable only to shooting by means of spring-guns. The answer to such an argument is to suggest that he who thinks such photography to be too easy should try for himself and see what luck he has. Simple though such photography may appear to be, there is far more in it than merely obtaining a tiger-kill, arranging the apparatus, enjoying a comfortable night's sleep in bed, and then arriving the following morning to find that the tiger has obligingly done the rest. I have kept careful records of all tiger-kills since I took up flashlight photography, and the following figures may be of interest. I have had fifty tiger-kills, of which eight have been natural kills and the remainder have been buffalo-baits. Yet I have obtained only eight good tiger negatives, or an average of one negative to over six kills, whereas sportsmen who shoot tigers by beating expect to get at least an average of one tiger for every four kills. In all, the flashlight was fired thirty-one times, in twelve cases by tigers and the remainder as follows:

Hyænas four times, wild cats, birds, pigs and jackals twice each, and sambar, leopards, porcupines and bears once each, whereas three times there has been no animal in the picture at all. Of the twelve times that tigers have fired the flashlight, twice the shutter has failed, on each occasion when the tiger happened to be in a particularly good pose, and I know nothing more annoying than to find a blank plate when the exciting and final act of development arrives, after months and months of weary effort have at last culminated in a flashlight exposure on a tiger. The remaining ten exposures have produced eight good negatives, but two of these were obtained at a considerable distance from the kill, so that fifty kills have really produced only six tiger negatives. Bearing these figures in mind, I do not think that anyone can justly claim that automatic photography is too easy to be sporting, when one has to compete with such heavy odds against its success. Indeed, many a time have I been on the point of giving it up altogether after a particularly long series of failures.

In the preceding paragraph I have stated that tigers have fired my flashlight twelve times in the neighbourhood of kills, but on other occasions also tigers have returned to the kills without firing the flashlight, either owing to the trip-wire breaking or else after the flashlight had already been fired by some other creature. The average tiger undoubtedly intends to return to his kill again after his first meal and many of the cases when tigers have abandoned kills altogether are directly due to putting the apparatus up too late in the day, to making too much noise, or to the firing of the flashlight by some other animal at a time when the tiger is near by listening or watching his kill. After all, most of the fifty tiger-kills in question have been in hilly country, where the tigers nearly always lie up in some elevated place, which gives them every opportunity of hearing or seeing any intruder upon their domain. With ordinary machan-shooting one can approach a kill very quietly, tie up the machan from the back of a tame elephant in a few minutes, and leave the place practically undisturbed ; but the erection of the flashlight apparatus takes anything up to two hours, during which time it is almost impossible to avoid making a considerable noise in erecting lamp-posts, camouflaging the camera with branches and blocking undesirable approaches to the kill, so that the tigers really have every justification in abandoning such kills altogether and one stands practically no chance with a really cunning old beast who has been taught wisdom by bitter experience.

Automatic photography can also be used for photographing tigers away from kills, but such photography is extremely uncertain and requires an intimate knowledge of the locality, as well as of the habits of the particular individuals one wishes to photograph. Tigers, when hunting at night, travel by preference along jungle roads, paths and dry river-beds, so that it is just occasionally possible to anticipate the exact spot which a particular tiger is likely to pass on a definite night. Most tigers have a certain beat, which may take them a fortnight or more to complete, and long observation tends to give one some indication of the average intervals at which the tiger is likely to pass certain spots. Incidentally, in order to obtain information on this subject, I have had maintained daily records of tracks passing definite fixed spots for 6 months or more and have thereby added to my knowledge of the movements of tigers and other denizens of the jungle. Having decided upon a suitable spot and a suitable time, one then arranges the apparatus in such a way that the tiger will fire it when passing, and hopes against hope that one's calculations are correct and that the tiger will arrive from the right direction before any other passer-by has had the time to spoil the chance by coming first. I have arranged my apparatus perhaps 150 times in this way, and I have had a tiger pass only six times. Four times the flashlight was fired, the fifth time the tiger saw the trip-wire and stepped carefully over it and the sixth time the wire broke without completing the electric circuit. Of the four times the flashlight was fired, once the shutter failed to work and another time the tiger found the trip-wire and took it in his mouth, the somewhat indifferent picture on plate xxx showing him actually in the act of biting it through! Of the remaining two exposures, one is figured on plate vii, and shows a tiger going away, and the other is pictured in the frontispiece and on plate xiv, and is, I consider, the best photograph I have ever obtained.

Editor's Note

1. *References to photographs not reproduced in this anthology have been retained for readers who may wish to look them up in the original book.*

A.A.D. BRANDER

TIGER: DISTRIBUTION, SIZE AND HABITS

The tiger is the most important animal in India; in fact, it is not going too far to say, that associated in the mind of the sportsman with the mention of the word India, is the thought that it is the land of tigers. No excuse is therefore needed for dealing with this animal more generally, and for not confining myself to its Provincial aspect.

Tigers are very widely distributed. They are found as far west as the Caucasus, in Northern Persia, in many parts of Central Asia, Manchuria, Mongolia, China, Korea, India, Assam, Burmah, and the Malay Peninsula as far south as the Island of Singapore. Colonel Count Nieroff of the Imperial Guard, St. Petersburg, who had done much big game shooting in many parts of the world, told me that tiger, which he had hunted, were much more generally distributed in Siberia and Central Asia than was generally supposed, and that they are often found in reed beds along the banks of rivers. The same might also be said of China, and I have been given a similar impression as to its distribution in this country by persons in a position to judge. The fact that tiger shooting is not regularly indulged in except in India tends to create the erroneous impression that Manchuria, India, Burmah and the Malay Peninsula have more or less a monopoly of this animal.

As is to be expected from an animal with so wide a distribution, tigers can accommodate themselves to a variety of conditions, and they are found in India in dense forest, scrub and grass land. Apart from food, the two essentials to their distribution are a sufficiency of water, and means of procuring shade from the sun, of which they are very intolerant. Except in Baluchistan, Sind, and parts of the Punjab and Rajputana, suitable conditions are met with throughout India, and save in such localities in which they have been shot out, tiger are still found all over the Peninsula, and for some distance into the Himalayas. Tigers however have not penetrated into Ceylon, and it would appear as if they had arrived in India after this island was separated from the mainland.

All tigers, wherever found, are essentially the same beast, but vary considerably in type according to the environment in which they live. Those coming from the Caucasus and Persia are generally a small hairy race. The Siberian and Manchurian tiger and those from the Amur are immense hairy animals, much larger than anything now found in India. In Korea such specimens as I saw, compared with the Indian beast, were higher in the leg, had a tendency to have withers, the neck appeared short and did not flow on from the shoulders as in the case of the Indian animal. There was also a depression in front of the shoulders. In size and weight however they were inferior to Indian tigers. As I only saw a few specimens I cannot say if these characteristics are general, but the Korean animal is essentially different from what one understands by the Manchurian tiger.

Tigers found in China and the Malay Peninsula have always struck me as being smaller and having a tendency to be darker than Indian tigers. In India itself, although individuals vary considerably in size and colour, there is no consistent variation in any particular locality. Animals from the south of the Peninsula probably average somewhat smaller.

While the Central Provinces produce animals fit to take their place with any in India, Sir John Hewett and others qualified to judge, consider that Bengal and Terai animals average somewhat larger. These areas are mostly covered with Sal forest, and it is significant that the tigers from the sal forests of the Central Provinces average larger than from the mixed forest areas. Those familiar with both have informed me that the Central Provinces animal is more truculent and will charge more readily.

* * *

Few questions have been more hotly debated, and the matter is by no means settled yet and probably never will be. Advocates of the 12 ft. tiger can quote numerous instances from books on shikar in support of their views, but this does not help much, as everything depends on the accuracy of the writer. That a great many were inaccurate is undeniable. Some of the sportsmen who hunted in the Forties of the last century refer to tigers of even over 12 ft. in the most off-hand way. One gets the impression that anything under 11 ft. is hardly worth talking about. In some cases these men were followed a few years afterwards over identically the same ground by others who record nothing but tigers of 9 and 10 ft., animals no bigger than those killed every year in India to-day. Amongst men of this class I may mention some mighty hunters with the lengths of the biggest tigers they killed.

Shakespeare	one of 10 ft. 8 in.
Simpson	two of 10 ft. 4 in.
Nightingale	one of 10 ft. 2 in.
Sanderson	one of 9 ft. 6 in.
Hamilton	one of 9 ft. 3 in.

What happened to *all* the big tigers? Did none survive? Did the first sportsman kill them all? Because we must undoubtedly reject many of these records, are therefore all to be similarly treated? That is the difficulty; to discriminate as to whose measurements are reliable.

Most of the authors whose style and records are most convincing give us nothing sensational. Eighty years ago when sportsmen first began to hunt tiger seriously, one must not forget that tigers were much more numerous than now, and often lived in country where no tigers now exist and which may have been specially favourable to their development. Amongst a great number of tigers the chances of an abnormally large animal occurring are increased. Moreover, their comparative immunity from serious hunting up to this time, gave the tigers every opportunity for the fullest development. In those days tigers were measured round the curves. In a normal tiger this adds 3 in. to 4 in. to the length; in a large tiger it may add 5 in., and the larger the tiger the more it tends to increase the length. To measure round the curves gives great opportunities for increasing the length, so much depends on how often the tape is pressed in.

Anyone familiar with the East, knows well how keen the native shikari and the babu are to 'please master', and we are never told exactly how these large measurements were taken. Even given four or five white men together and getting each in turn to measure round the curves their returns will all vary and some by as much as 3 in. The proper way to measure a tiger is to place it on its back on a flat surface and depress the head, then place two uprights, one touching its nose and the other the tip of the tail—remove the carcass and measure the space between the base of the two uprights. Measurements made in this way will not vary by ¼ in. It is only comparatively recently that sportsmen have commenced to measure in this way. Allowing for the extra 5 in. owing to past methods of measurement, although this will bring the 11 ft. tiger down to within reasonable distance of animals still occasionally killed, it utterly fails to account for the 12 or 12 ft. 4 in. beast.

I know of one authentic case of a tigress measuring 9 ft. 6 in. between uprights, and assuming this animal to mate with a very large tiger, and the cubs to have a prosperous upbringing, and taking the Indian tiger's ancestry into consideration, it is impossible to say that a tiger cannot exceed 11 ft. Nevertheless at the present time the announcement of the shooting of an 11 ft. tiger would be followed by a deluge of questions and since the interest of the whole sporting world has been centred on 11 ft. tigers these have ceased to materialize.

The more improbable the event, the stronger the evidence demanded in order to establish the same. It is significant that Jerdon, in his *Mammals of India*, published in 1874, will not accept even the 11 ft. tiger, and he must have known men who laid claim to having killed animals of over this length. While wishing that fuller details had been furnished as to how and by whom the measurements were made in past days, I think we can accept the occasional tiger of 11 ft. measured round curves, but we certainly cannot extend indulgence beyond this. It is hoped that anyone in future killing a phenomenal tiger will furnish the fullest details regarding the measurements. This is specially required in cases of large animals being killed by important personages, to please whom many are only too ready to stretch a point, and such persons are apt to delegate the work of measuring; at any rate they are always surrounded by a staff to relieve them of personal supervision and they are largely immune from the pertinent enquiries of the sceptic.

The weights given in old records are on the other hand often immoderately small, so much so that when compared with the stated length, and assuming that both are correct, the modern tiger is an entirely different animal. Some of these records have been published in the *Bombay Natural History Society's Journal.* In some cases tigers of the largest size are only made to weigh as much as a fair-sized tigress, and it is impossible that both weight and length are correct. Sir John Hewett probably possesses the most extensive records of any living person, covering as they do 241 animals which he has seen shot, mostly in the Terai. He has been kind enough to favour me with some of his figures. The longest tiger he ever saw shot was 10 ft. 5½ in. and the longest tigress 9 ft. 6 in. The heaviest tiger he weighed was 570 lb. and the heaviest tigress 347 lb. Out of the 241 animals nine tigers were 10 ft. or over and ten tigresses were 9 ft. or over 9 ft. One of these was from the Central Provinces.

An exact comparison with animals from the Central Provinces is not possible as the measurements recorded by me were taken between pegs and this reduces the length by 3 in. to 5 in. Another point is that I weighed and measured very few immature animals. The figures I possess refer to forty-two fully mature males and thirty-nine fully mature females, and while I have seen many more mature animals killed than this, I have not got a record of these, and it is probable that in selecting animals for measurement the tendency would be to choose large specimens. Unfortunately some of my diaries were lost or stolen in Bombay. I am unable to say, therefore, exactly how many tigers I have seen shot, but it can be taken as being approximately 200. Out of this number one is of 10 ft. 3 in. and another of 10 ft. 2 in. In addition, there are one of 9 ft. 11 in., one of 9 ft. 10 in., the latter shot by His Royal Highness the Duke of Connaught at Supkar. Another tiger at the same shoot was 9 ft. 9 in. I have another record of a tiger 9 ft. 10½ in. All these six animals, if measured round the curves, would have been 10 ft. or over. The biggest tigress I have seen measured was 9 ft. 1 in. In addition, I have records of two of 8 ft. 11 in., one of 8 ft. 10 in., and one of 8 ft. 9 in., i.e., five animals of 9 ft. or over, if measured round curves. It would seem, therefore, that the occurrence of 10 ft. tigers and 9 ft. tigresses is slightly more common in the Terai and Nepal than in the Central Provinces.

The classification of what is a mature animal has presented some difficulty, and would vary according to the views of the individual. Out of the thirty-nine tigresses selected as mature, the smallest was 7 ft. 10 in and the largest 9 ft. 1 in. The average is 8 ft. 4 in. and the average weight is 290 lb. The heaviest tigress weighed was 343 lb. The shortest tiger classified as mature was 8 ft. 9 in., and the longest 10 ft. 3 in. The average works out at 9 ft. 3 in. The weights vary between 353 and 515 lb., averaging 420 lb. for a gorged tiger. I weighed one other animal over 500 lb.

The largest animal I actually ever saw, however, only taped 9 ft. 11 in. between uprights. Most unfortunately, I had no weighing machine and could only make a rough estimate of its weight by balancing it against a number of men, and some of these men left before I could weigh them. I firmly believe this tiger was about 600 lb. in weight. This animal had been living on full-grown buffaloes, and was doing an immense amount of damage. It had killed some of the largest animals, including a bull, with apparent ease. The bulk of its neck may be gauged from the fact that while biting its fore paw, after the first shot, I put in what I hoped to be a 'finisher' from a Ross rifle, using the usual copper-nosed bullets. The bullet never even reached the bone, and only stirred the beast up. The line of the backbone was sunk in a depression, and on each side the flesh came out as flat as a table for 8 in. before the curvature of the ribs commenced. In addition, it had what I have never seen on any other tiger. There was a curious wedge, not soft and flabby as one sees in Zoo animals, but a hard firm ridge 4 in. to 5 in. deep running all along the belly from the pelvis to the throat. It was so distinct I thought it must be due to some disease, but it turned out to be merely a strip of pure white fat. Some idea of this animal's size can be gathered from the following measurements:

Length of body	7 ft. 3 in.
Length of tail	2 ft. 8 in.
Girth of body	59 in.
Girth of head	39 in.
Girth of forearm	21 in.
Height	43 in.

The biggest tigress, 9 ft. 1 in., already mentioned, was an old beast and was killed in the same jungle as the above-mentioned monster and may have been his mother. In both these animals the tail was short.

Although the size of tigers varies considerably, the shape does not. The greatest difference is found in the tail, which may vary as much as 15 in., and the length is not dependent on the size of the animal; the longest and shortest tails I have measured were 45 in. and 30 in. respectively.

The height of tigers at the shoulder varies from 36 to 44 in. A good average male should measure 39 in. Other average measurements are 37 in. round the head, 32 in. round the neck, and 19 in. round the forearm.

It will be seen from these measurements that there is nothing lanky about the tiger, and that he is a large, powerful, burly animal, differing entirely in a wild state from the impression one gets from looking at specimens in captivity. In fact, I have seen quarters on a tiger that would not have shamed a polo pony. Tigresses are, of course, much smaller than tigers and generally more sinuous. The appearance of the tiger is too well known to require general description. It will suffice to mention some points which may be overlooked, and a few variations from type.

Lydekker merely mentions that they have 'a short ruff on the throat'. In my opinion, this statement is inadequate and hardly correct. In their winter coats, old males often have regular 'Dundreary whiskers' in the region of the chops immediately below the ear and all have a certain amount of 'ruff' on the top of the neck and round the throat which in cold climates develops into a small mane. In summer the hair is much shorter and paler in colour. In some specimens the white of the belly extends some way up the ribs, thus resembling a colouration more commonly seen in Northern varieties.

All tigers are born with their full number of stripes, and these persist through life, but in old specimens they tend to become fainter. The number of stripes on tigers and tigresses are practically the same, no sex having consistently more than the other. Young tigers and tigresses give the impression of being more striped, but this is merely due to the body being smaller. The impression that some stripes may have disappeared from the shoulders of a heavy old male is due to the development of the muscle and the larger space between the stripes. As one so often sees mounted specimens with pink insides to their mouths, it may be as well to mention that the insides of tigers' mouths are a light whitish olive green, and the tongues are yellowish white or grey but faintly tinged with pink. White tigers occasionally occur. There is a regular breed of these animals in the neighbourhood of Amarkantak at the junction of the Rewa State and the Mandla and Bilaspur districts. When I was last in Mandla in 1919, a white tigress and two three-parts-grown white cubs existed. In 1915 a male was trapped by the Rewa State and kept in confinement. An excellent description of this animal by Mr Scott of the Indian Police, has been published in Vol. XXVII, No. 47, of the *Bombay Natural History Society's Journal.*

Blanford, in his *Mammalia*, mentions the occurrence of a black tiger in Chittagong, and this statement is repeated by Lydekker. One cannot help wishing that the evidence on which this rests had been more fully stated. A large black panther might have been mistaken for a tiger. Col. Burton, in an excellent article on Panthers in Vol. XXVI, No. 1, of the *Bombay Natural History Society's Journal*, discusses black tigers, and accepts this instance as 'proven' but rejects other cases. Personally, I have always felt sceptical about this animal. One occasionally hears stories of black tigers, but they are never shot. One cause giving rise to these tales might be that the tiger had been rolling in the black sooty ashes that often result from a jungle fire. Another false impression can be conveyed by caked blood on the tiger.

I once watched three tigers feeding on a fresh kill, and the largest animal which had of course selected the favourite place between the buttocks, managed to get itself smothered in blood, all the visible white being covered. As I was watching this performance, which was in broad daylight, the red of the blood changed to black as it rapidly does, and had I not witnessed this transformation and come on the tigers without being aware of what had happened, I would have been firmly convinced that I had seen a black tiger.

* * *

On one occasion I was out in a Sal forest at daylight. The cries of deer and monkeys advertised the presence of a tiger or a leopard in a neighbouring maidan, so we pushed the elephant along and came on a very large tiger crossing the bare open ground of a deserted field. Two barasingha stags were standing at the edge of the field and within 12 ft. of a wall of grass 10 ft. high. They were close together facing the tiger, but braced back, and stood motionless, braying. The tiger's line would have taken him past the stags, but he gradually swung in and approached the stags which allowed him to come within 20 yards. The tiger then straightened his tail, lifted one huge fore paw, and started the usual preliminary swinging which indicates his intention to charge. The stags seemed rooted to the spot and had by this time ceased to bray. They were certainly in the greatest danger. This seemed to be entirely due to their inability to move, as long before danger arose, they could have disappeared at one bound in the long grass. I was halted at the edge of the forest about 80 yards off. Unfortunately, the play ended before the last act. I was accompanied by a friend who was anxious to shoot the tiger. He was using a heavy rifle and black powder. Everything vanished behind a cloud of smoke and when this cleared the maidan was empty.

Man is so feeble an animal, tigers do not require special methods of attack in order to break his neck, and can kill him easily by biting him either through the body or the neck. I once saw a case in which a man-eating tiger had driven in a man's skull with a blow from his paw delivered on the top of the head. Evidence of how man-eaters behave in killing, or towards a live victim which they can carry off as easily as a dog does a rabbit, is of course scanty. Apropos of their strength I once saw a tigress drag a half-grown buffalo up the bank of the River Tapti without apparent effort. The bank was alluvial soil and so steep, a man could only have climbed it with difficulty without the use of his hands.

I was once stalked and charged by a man-eater which got within a few feet of me before I shot it. I cannot say, therefore, if it would have taken me at the gallop, or pulled up and then bitten, but this is probably what it would have done. Man-eaters usually reject the skull and the palms of the hands and the soles of the feet. The story of Jezebel was evidently recorded by an accurate eye-witness, as animals which eat men seem to dislike these parts.

When a wounded tiger charges, if the man is running away or moving, they will take him at the gallop, but if one stands still they invariably pull up dead and then attack. But for this habit, some sportsmen, including myself, would not be here today. In attacks by wounded tiger they usually knock the man down and give him two or three rapid bites, often through the thigh, and then leave him. I have known an unwounded tiger in a beat, however, round on the men, and after mauling one, pick up another and carry him off. In this case, although the tigers had been headed back, the attack was chiefly due to the tiger being with a female at mating time, and after the tragedy he came out on a fireline dancing on his toes, with his tail waving in the air and evidently very proud of his prowess, for which he no doubt expected to be rewarded.

Tigers, when making a galloping charge at man, give vent to a deep short grunting cough repeated two or three times. The volume of sound is so great, it is sometimes difficult to locate its direction accurately. On occasions they will steal in and attack silently, and they nearly always attack animals in silence. A tiger suddenly disturbed and retreating often gives a 'whoof', and when trying to break a line of beaters they will gallop along the line roaring. This noise is quite different, however, from the cough they make when intending to attack. Tigers growl as a warning when angry. A deep long drawn out 'meow' is another call, and when pleased they make a purring noise by blowing air on to their lips which vibrate, not unlike the action and noise made by horses when they wish to clear their mouths and nostrils. They show emotion by a twitching in the end of the tail. This may be either pleasure or annoyance.

Cubs make complaining squeaks, but on the whole the tiger is a silent animal, and considering their numbers they are seldom heard to give a regular roar. When they do this the sound is awe-inspiring, and the whole valley will sometimes ring with the volume of sound produced. The noise is produced by taking a deep inhalation and then expelling the air violently against the roof of the mouth, the lower jaw being slowly closed at the same time. The result is a long drawn out 'H—o—w—n'. They sometimes repeat the sound three or four times in succession, and commence moderately, but 'work up' and with each repetition increase the sound. When tigers do this they are generally trying to get in touch with a mate.

On one occasion, when camped in the middle of a forest, I heard the roaring of a tiger taken up and answered by another. The two animals gradually approached each other and a terrific fight ensued not far from the camp, the scene of which I was able to visit next day. About six weeks after this, in all probability I shot one of these tigers, judging by the marks on his neck and body. He was an immense beast. Nevertheless, he appeared to have had the worst of it. However, we never saw 'the other fellow', and this deduction may have been that of 'the casual observer'. For some unknown reason tigers are more given to roaring in some districts than in others, and I have always found them more noisy in Sal forests than elsewhere.

One more noise, which is rarely heard outside Sal forests, has to be described. This is the curious 'pook' they make not unlike the sound produced by a sambar, and about which there has been considerable discussion. This noise can be mistaken for that of a sambar, but it is really softer and not so loud or harsh. No sambar could mistake it for the call of another sambar, and the suggestion that tiger call up sambar in this way can be turned down. Moreover, the noise made resembles the sambar's cry of alarm, and how this could be an inducement for the sambar to approach the spot whence the alarm issues, is not understood. It might be argued that by making the noise the tiger induces a sambar to 'bell', and thus locates him, but the tiger has other means of doing this. The call is really a mate call and is used by tigers to locate each other. I have nearly always heard it made when there were two tigers going about together. I have had a tiger make it at me, thinking I was his mate or at any rate wishing to find this out.

Two tigers had killed a fairly large buffalo in long grass. The tigress had left the kill, but when I came to it the tiger was on the buffalo. Hearing me coming he picked up the buffalo and took it off with him through the grass which was very high and dense. I followed up, when he dropped the kill, and commenced this noise at me. He repeated it a number of times from different points and retreated making the noise. Of course, I could not give him the reply he wanted, but this came presently from the hill-side, and he went off and joined his mate. These tigers both returned after dark and spotting me in a tree roared lustily for some hours. Seeing it was useless to wait longer, I got down and returned to camp. They then came up to the kill and ate everything except the skull and hoofs.

Tigers seldom attempt to climb trees, but that they can do this to a much greater extent than is generally supposed, I firmly believe. They can certainly get up a branched tree at which they can rush. The number of fatal accidents is evidence of this. I do not think a large heavy tiger could climb a smooth limbless tree, but tigresses and smaller tigers can get up a tree with only a small amount of assistance from side branches. Credible witnesses have told me on different occasions that they have seen tigers treed by wild-dogs, and I have been shown the hair of a tiger on a branch 15 ft. from the ground.

Tigers take to water readily and are strong swimmers. I have known them swim the Nerbudda to escape out of a beat, and on one occasion a wounded tiger that was being hunted did this. They must have swum to get on to the Island of Singapore, and some years ago it was reported that a tiger had actually crossed over on to Hong Kong. They are great wanderers, and at times seem to lose themselves and get into the most extraordinary places. There is a circumstantial account of one being found in a pagoda in Burmah. Three tigers were once found in the Buldana District trekking across open country, and miles from any real holding ground.

Again, while I was Divisional Officer of Melghat, a young tiger left the jungles and took up its abode in some Pan gardens near Ellichpur. It was hunted and actually shot in a Pan garden after an exciting chase in which it displayed the activity of tigers. I was informed that, during the hunt, the tiger leaped the sides of the garden, which were 6 ft. high, like a greyhound. I myself have seen a tigress clear a 19 ft. gully in one stride without effort or gathering herself to leap, and in alighting she gave the impression that she would not have crushed an egg, so easily was her weight carried. Straying into unusual places is the more surprising as the chief characteristic of the tiger is his retiring habits. Their chief endeavour in life seems to be to avoid being seen or having attention drawn to them. On the other hand, there are instances of extraordinarily bold behaviour at night, and they have been known to return to their kills again and again after being fired at. These cases however are exceptions.

For so large an animal, tigers must be considered rapid breeders. At one time in parts of India at the beginning of the last century, they were so numerous it seemed to be a question as to whether man or tiger would survive.

Up till about the beginning of the present century, sportsmen only visited the Central Provinces in moderate numbers, but about this time shooting became a popular pastime amongst army officers, and tigers were much reduced. The war practically put an end to shooting, except by district officers, and during its duration, the tigers rapidly increased. Few tigers are killed by native shikaris. These chiefly shoot deer. It is the European sportsman that thins out the tiger.

Any cause which prevents a tiger getting his natural food, tends to create man-eaters, as tigers fall back on the easiest animal of all to kill and his instinctive dread of man is overcome by hunger. Moreover, this instinctive fear soon goes once he realizes how helpless man is. Tigers may be unable to procure their ordinary food by reason of worn-down teeth, lameness, or some wound. Again, there may not be a sufficiency of natural food to meet their requirements. A man-eating tigress will also bring up her cubs to kill man. Since the war, there has been a very large increase in the number of man-eaters. Hardly a gazette is issued without announcing special rewards for about twenty different animals. The bulk of these animals is in the east of the Province. It is possible that the rapid increase in the number of tigers during the war, with no increase in the food supply, but rather the reverse, has led to the present conditions.

The presence of a confirmed man-eater in a jungle tract is a dreadful scourge. Some people have to pass through jungle almost daily, and no man on leaving his village can be sure of returning. The man-eaters often develop boldness and a fiendish cunning, and their presence is a blight on the whole community. A book could be filled with jungle tragedies.

The knowledge man-eaters acquire of men's habits, as well as their habit of eating up most of the corpse at a meal and not returning, adds to the difficulty in killing them. Moreover, there is a widespread belief that the ghost of the last victim sits on the tiger's head, and that he who gives information regarding the tiger's habits is selected by the ghost for the next victim. There is often much difficulty, therefore, in obtaining the necessary assistance to encompass a man-eater's destruction. If the perusal of this chapter induces even one sportsman to hunt down one of these man-eaters, it will not have been written in vain.

Animals which have become confirmed man-eaters usually decline to kill a buffalo or bullock, but will often kill a pig or a pony, if these are tied up as baits. I have known other tigers, not man-eaters, to have a similar preference, but the disinclination of man-eaters to eat beef indicates that their palate has undergone some definite change.

The great mass of tigers are partly cattle killers and partly game eaters, but some animals living in forests such as are found in Mandla and Balaghat, where game is very abundant, hardly ever kill cattle. On the other hand, many parts of the Province now contain little or no game, and in such places the tiger lives on nothing but cattle, and follows the herds about as these are shifted for water and grazing. In tracts of this nature tigers are particularly liable to develop into man-eaters.

In the Zemindaris of Bilaspur, I found the tigers had definitely changed their habits and did all their hunting by day. It was some time before I found this out, and I was surprised at the difficulty in getting a bait killed, which of course was tied out all night. The facts were that there was absolutely nothing for a tiger to hunt at night, as there was no game, and all the cattle were driven into pens at dusk. Each tiger made a round of about six villages, taking a bullock from each in turn. On the other hand, I have known tigers living in forest containing little game and yet seldom killing cattle, and it has often been a matter of no little speculation when shooting an animal in the pink of condition in jungles of this nature, where the tiger obtained his food.

Although tigers do a great deal of damage, they have their uses in preserving the balance of nature. There is an outlying patch of forest in the Hoshangabad District which always contained a few tigers when I was there in 1906. Some years after this, they were all shot out, and the forest being isolated no others wandered in. I visited this tract again in 1917, and the surrounding villages were simply overrun with pig and nilgai. Many fields had gone out of cultivation. To enlarge a couple of tiger in this forest would be a great boon to the local people. This is not the only instance of the kind, and the extermination of tiger in such places should not be permitted.

KENNETH ANDERSON

THE SWAMI OF VALAITHOTHU

Most of you who read these stories and have been to India will know that the Tamil word *swami*, signifying awe of God as the Supreme being of the Universe, is applied to any of the many deities in the Hindu pantheon, and used also to describe a holy or devout man. In its more colloquial use it is merely a term of respect for a recognized gentleman. *Valaithothu* is another word from the same language, and in the context of this story applies to 'a banana plantation'.

But the 'Swami of Valaithothu' was neither god nor man in a banana plantation. He was a tiger that lived in the foothills of the Nilgiri Mountains, and haunted the road, the firelines and the jungle stream that skirted a large grove of bananas just twelve miles from Ootacamund.

His claim to this seemingly inappropriate appellation came about through a chain of rather strange and apparently unconnected circumstances that began almost a decade earlier.

For at that time a sadhu, or fakir, wandered down the road one evening, presented himself at the grass hut in which the owner of the banana plantation was living and demanded shelter. He said that he was on a journey on foot through India that had begun in the 'abode of snow', the Himalayas, and was to end at the southernmost tip of the peninsula. He also said he would stay for a day or two to rest, before climbing the Nilgiris to Ootacamund.

The banana-grower, whose name was Kantha, was both honoured and delighted. He thought that this visit from the holy man would bring him and his plantation good luck and fertility. He suggested to the fakir that he should remain a week at least, to rest his weary body, before attempting the ascent of the steep ghat road that lay ahead, and he coaxed his visitor with a large dishful of delicious bananas, spread with thick fresh cream, liberally sprinkled with sugar.

The itinerant holy one enjoyed the fruit and cream so much that he remained not only for the week for which he had been invited, but for many weeks after that. In short, he became a permanent resident and promised the banana-grower his blessings and great productivity if a nice hut were built for him, and the bananas and cream and sugar were made a constant item on his daily menu, to be served fresh, early each morning.

The owner agreed for fear of reprisals if he did not, and the sadhu settled down in his new grass hut to a life of ease, meditation and good fresh fruit and cream.

It did not take long for the few Irila tribesmen, working on the plantation, to observe that the newly-arrived holy man was a queer sort of customer. After cooking and eating his evening meal, he would retire early into his hut, and the faint flickering flame of an oil-lamp, visible through the interlaced bamboos that formed the walls of his abode, would burn late into the night and sometimes all night long.

On one occasion, after the holy man had been seen to enter, one busybody prompted by curiosity, crept up to the walls and ventured to place a curious eye to an aperture to find out what was going on within. But the hut was empty!

The spy, thereupon presuming that the fakir must have gone out again unnoticed, and hoping to catch him red-handed on his return, hastily made himself scarce among the closely-growing stems of the banana plants, and at a safe distance squatted on the ground to await the fakir's coming and see what he could see.

A few minutes passed, and then a tiger roared loudly in the jungle just beyond the sadhu's hut.

The watcher started up in terror and turned to beat a hasty retreat to his own quarters. But fear caused him to turn round, half-expecting to see the tiger in hot pursuit, crossing the little clearing that separated the bananas from the forest; but to his utter amazement he saw the sadhu walk boldly forth from the very direction where the tiger had been roaring.

Hastening back to the cooly-lines, the spy told his story. What sort of man was this, the Irilas wondered, who consorted with tigers when night had fallen?

Thereafter, they noticed that at night a tiger roared in the vicinity of the sadhu's hut, and indeed all around the plantation, very frequently. From the timbre of his call, the animal was obviously a male. Other tigers there were in plenty in the surrounding jungle, both male and female, and their calls could also be heard at infrequent intervals, generally far away.

But this particular beast was clearly a newcomer to the region and appeared to have taken up his abode in the jungle just beyond the limits of the banana-grove. They frequently found his fresh tracks in the morning. They were everywhere: upon and beside the road coming down from Ootacamund, along the banks of the stream, all over the clearing that ran as a rough rectangle around the plantation, separating it from the forest, even in the moist earth among the bananas, and very frequently around the fakir's hut.

The tiger called nearly every night, particularly on the occasion of the *amavasa*, or new moon, and people were afraid to step out of doors once darkness had fallen. But for all these evidences of the tiger's close and permanent presence, nobody ever caught a glimpse of him during the day.

The forest was full of coolies at the time; some were engaged in bamboo-felling; some in gathering the ripe fruit of the tamarind trees, and some on road repair work. Apart from the coolies, there were a number of Irila herdsmen quartered at different cattle-patties scattered throughout the area, employed to graze large herds of cattle and buffaloes belonging to rich Badaga landowners at Ootacamund, who had purchased grazing-licences and sent their herds down to the jungle for easy and economical feeding. These coolies were out all day with the animals, grazing them in the luscious spear and bison grasses that everywhere clothe the foothills of the Nilgiris. On all sides they found the huge, saucer-sized pug-marks of the lone male tiger that called so loudly and frequently almost every night and they lived in daily fear of an attack on their herds which they knew was inevitable at any time in the near future.

But the attack did not come. At least it never came during the hours of daylight, as might have been expected, when the cattle were out in the jungle. For then they were most vulnerable, scattered as they were in the thick mass of shrub, grass, trees and bamboos. Instead, the tiger chose the hard way to take them. He did so at night, when they were supposedly safe in their pens.

The walls of these pens consisted of a close array of wooden stakes, driven in the ground and standing above it to a height of perhaps five feet, reinforced along the outside with bundles of thorns cut from the jungle.

One morning Boodia, one of the Irilas, found a fine, fat, brown milking-cow missing from among the herd entrusted to his care. The wooden palisade was quite intact, but on the soft earth around it were the large pug-marks of the solitary male tiger, showing that he had walked around the stockade, had deliberately selected a place where the stockade was a few inches lower, had jumped over it, killed the cow and jumped back at the same spot, carrying his victim with him. Clear evidence for these assumptions was found at the place inside the stockade where the tiger had killed the cow: her flailing hooves had scored the earth and there was blood on the ground that had seeped from her throat. Also, the top of the stockade had been broken in an outwards direction, when the body of the dead cow had fouled the stakes as the tiger had leaped over with his burden, causing him to use force to disentangle it and bring it over with him. The deep indentations on the earth outside indicated where the tiger and his victim had finally landed.

No time was lost by Boodia in telling the other herdsmen of what had happened, and that day was a busy one for all the graziers, who devoted all the available hours of daylight to raising the height of the stockades around their respective patties.

This was of course a considerable undertaking, as each pen had no more than two or three Irilas to look after it. One of these had to continue to graze the cattle, while the second man—and perhaps another to help him—cut the new stakes to size, carried them to the stockade, and drove them into the earth within the existing row. There was a lot of quarrelling and argument, because every man preferred the easier job of grazing the herd.

By the time all the pens in the area had been fitted with the new, higher and heavier palisade, the unwelcome tiger that had come to stay had accounted for four more cattle from the various herds, all of which he took by night and in the same way as he had stolen the first. Buffaloes he did not touch, evidently because they were too heavy to be lifted over the wooden fencing.

These activities created a furore among the Irilas, for never in the memory of the oldest among them had any tiger ever behaved in such a fashion. Having lived for generations in the forest, they were familiar with all the ways and habits of the carnivore. Marauding tigers and panthers had every year taken their toll from among the herds, and at very infrequent intervals a man-eater had appeared on the scene. Then terror had been their constant companion till such time as, in one way or another, the reign of the man-eater had been brought to an end. But the tigers that had killed their cattle had all struck during daylight, when the herds had been taken out to graze. The occasional man-eaters had fallen upon their victims along the lonely tracks or when they were gathering jungle fruit, but these attacks had always been launched by day.

Here, however, was a tiger that apparently moved only under cover of darkness. They heard him at all hours of the night, and they saw his pug-marks each morning, but nobody had ever seen him. The five head of cattle that had been lifted over the palisades at night proved his existence to be very real indeed.

Why did he not launch his attacks during the hours of daylight, like any other tiger? And with that came the awful thought: if this tiger ever became a man-eater, their lives would indeed be in peril. His assaults and his method of removing his victims proved him to be an extremely cunning animal, and quite fearless.

While this was going on, tongues had been wagging. Of course, they only spoke in whispers and after many glances in all directions to ensure there were no listeners. Was the strange sadhu on friendly terms with this wily tiger? Was it possibly he who secretly met the animal and instilled into him the cunning that had led him to prey upon the herds in such an unusual fashion?

Quite soon all the cattle-patties in the area, right up to the border of Mysore State, over fifteen miles away, had had their palisades raised to a height averaging ten feet or more, and the herdsmen anxiously waited to see what the lone tiger would do to overcome the problem that now confronted him.

Then something happened that instilled even greater fear into the superstitious minds of the Irilas.

About that time a wild bull-elephant had made one or two incursions into the banana-grove, spreading destruction in all directions. Although the shooting of elephants is strictly prohibited, Kantha, the owner of the bananas, decided to make at least an attempt to protect his orchard by firing a handful of miscellaneous lead pieces at the elephant through his old muzzle-loading gun when the beast made its next raid. Being moonlight at the time, he would be able to see the pachyderm and from a safe distance fire a shot to drive it away from his bananas, and possibly deter it from coming back.

Accordingly, Kantha concealed himself among his trees in the far corner of the plantation that the elephant had been raiding, and awaited the coming of the ponderous robber.

But no elephant came that night. Instead, the lone tiger roared nearby, and then roared again. A little later, in the bright light of the moon, a slight movement caught Kantha's eye in the open space separating the boundary of his land from the forest. He stared hard; and there, sure enough, was the stealthy form of an animal emerging from the tangle of undergrowth.

Kantha looked again. There was no doubt about it. Before him was the slinking form of a tiger—the tiger that had just roared—the lone male tiger that had been causing all the mischief and which nobody but himself had seen so far. What an opportunity lay at hand to raise himself sky-high in the esteem of the Irilas!

The distance was rather long for a shot with a muzzle-loading gun. If only he had a rifle. Nevertheless, taking aim as best he could in the moonlight and allowing for the movement of the target, the owner of the plantation fired what he hoped would be the shot that would turn him into a hero.

The shot did have far-reaching consequences, but whether it enhanced his reputation or lowered it is a debatable point.

That at least one or two of the irregular bits of lead vomited forth by his old weapon that night found their mark is certain, for the tiger let out a loud roar, followed by what the shocked man later described as an eerie scream, and vanished with a bound. Needless to say, without further ado, Kantha hurried to his hut and secured the door tightly.

The next morning, when the owner's son, a lad of about seventeen, brought the usual daily offering of bananas and cream to the sadhu, he was surprised to find the door of the hut closed on the inside. Upon the threshold, and leading to it, were drops of a rusty brown colour. They had soaked into the earth, but the boy had no difficulty in recognising them as blood!

Timidly, he called to the old hermit. At first there was no answer. Then a faint voice told him to go away. The boy replied, saying he had brought the fruit and cream, and the fakir answered that he was ill and did not want anything to eat that day.

His curiosity aroused, the lad asked if he could be of any help, mentioning that he saw blood on the doorstep and inquiring if the sadhu had been injured. The only reply to this was words of fury, 'Get out! Get out!'

Wondering at this strangely cold and ungrateful response to his well-intentioned offer, the boy hastened back to tell his parents.

Fearing the outcome of his shot at the tiger the previous night, Kantha had told no one of what had happened except his wife. Now his son's words caused him to think, and to think hard!

Half-an-hour's intense concentration on the problem resulted in what he considered a clever plan to find out if the grave suspicion dawning in his mind had any basis in fact. He hurried to the fakir's hut, approaching the entrance silently on bare feet. Gently he pushed against it, noting as he did so the dried drops of blood on the ground.

The door was fast secured from within.

Taking a couple of paces to the right, he dropped to his knees and pressed one of his eyes close to what looked like a sufficiently-wide aperture between the interlacing bamboos to permit him to see inside the darkened hut.

Just then a faint noise to his left and above, caused him to look up. The door of the hut had opened. Gazing down on him, with a horrible scowl on his face, stood the mysterious sadhu. His left arm, around the bicep, was roughly bound with cloth, and seeping through at one spot was the rusty-brown stain of dried blood.

Kantha rose guiltily to his feet and stammered: 'I heard from my son that you were ill, most holy one, and came to see if I could be of any assistance.'

'Does that explain why you were peeping into the hut, O liar-who-cannot-lie-properly?' thundered the fakir. Then, 'Go away, before something unpleasant happens to you,' he hissed, slamming the door in the astonished Kantha's face.

Rising hastily to his feet, the banana-grower beat a quick retreat, thoughts of black magic and the hermit's dire revenge springing into his mind.

A week or so later, the lone tiger took his next cow from one of the patties. As I have said, the palisades had been raised so that it was now impossible for any tiger to leap over them into the enclosures, and doubly impossible to leap back into the forest again, burdened with the weight of a victim.

So the strange tiger did the next obvious thing. He forced his way through at the base of the stakes, by displacing a couple of them with his powerful paws. But in this case he was unable to return with his prey after he had killed it. The tiger tried to do this, but the carcass became wedged between the stakes and he was forced to abandon it.

Thereafter the tiger became bolder, as the Irilas living in the area became more timid. At this stage in the developments, apparently through complete disregard for the human beings with whom he came into contact, and recognizing the fear he instilled into them, the tiger changed his tactics. Instead of operating only after nightfall as he had been doing, he was now frequently seen during daylight, walking along one or other bank of the stream or the fire-line that ran close to the plantation. Growing increasingly defiant, he began to pounce upon the herds of cattle that the Irilas drove into the jungle to graze.

The next two stages in this animal's misdemeanours were inevitable. As a rule the herdsmen, who closely associated the lone tiger's existence with that of the swami living at Valaithothu, gave the plantation a very wide berth, nor did they attempt to go to the aid of any of their stricken herds during an attack.

But there is always the exception to every rule. The tiger leaped upon a buffalo one day in full view of a herd-boy who was grazing the animals. The buffalo plunged about in an attempt to dislodge his attacker, while the other buffaloes, as buffaloes often do, milled around in confusion, trying to aid their stricken companion.

Taking courage from the efforts of his charges, the herd-boy rushed in among them with his staff and poked it directly into the face of the tiger. With a snarl, the tiger left the buffalo and leaped upon the boy, who crumpled to earth with his attacker on top.

Fortunately for him, that gave the other buffaloes the chance they were waiting for, and like a squadron of cavalry they charged the tiger, now fully exposed and within range of their formidable horns.

The tiger did not stay to fight. He was off and away, with the line of buffaloes charging in his wake. By their action they had temporarily saved the life of the herd-boy, although, by trampling over him, they had added their own quota to the injuries he had already received from the teeth and claws of the tiger.

All that night the luckless boy lay in the jungle with none to succour him, but his buffaloes kept watch around him against the tiger's return. Next morning a few of the neighbouring herdsmen banded together and started a search. They found the boy at last, but his multiple injuries and the many hours of exposure he had endured exacted their toll. He died before a conveyance could be pressed into service to carry him up the hill to the hospital at Ootacamund.

A long spell of comparative quiet followed before the next tragedy. There had been a lull in the attacks upon the various herds; in fact, many people thought the tiger had of his own choice moved to some other sector of the forest.

But one morning an Irila saw a pea-hen suddenly launch itself into the air from behind a large tuft of grass bordering the footpath along which he was walking and flap heavily away. His mind immediately associated this action with the presence of a nest and possibly a few eggs or even some pea-chickens to eat, and the Irila hastened towards the spot the hen had left.

But the cause of the alarm that had frightened the bird was the lone tiger, and he and the Irila met face to face just beyond the tuft of grass.

For the tiger it was a most annoying surprise. For the Irila, the last one on earth. For no accountable reason the tiger struck him down and killed him. Then, perhaps as an after-thought or through mere curiosity, he ate a small portion of the man's chest and gnawed a thigh.

The tiger liked the taste of the flesh and so that day another man-eater came into existence.

The consternation that had been growing steadily with the lone tiger's increasing boldness now turned into panic. Idle toungers driven by fear had been talking, and the feeling became prevalent that the swami living at Valaithothu was closely connected with the latest incident and with the tiger himself.

As regards the latter, there were divided opinions. A number held that the fakir and the lone tiger that had turned into a man-eater were on very close terms. Others felt that the man and the animal were one and the same; the sadhu being able to take on the form of a tiger whenever he wished. Such an explanation seemed to account for all the mysterious events that had taken place; the coming of the tiger that had coincided with the arrival of the sadhu; the wounding of the animal by Kantha and the drops of blood before the fakir's hut, and the man's bloodstained, bandaged arm; the tiger's frequent calls and pug-marks in the vicinity of the fakir's hut; his earlier clever and unusual methods of robbing cattle from the patties; and now the killing and partial eating of the Irila.

Incidentally, was it only a coincidence that the Irila who had met with this shocking and apparently quite uncalled-for fate happened to be the very person who had first pried upon the fakir and had started the rumours connecting him with the tiger? Surely, the sadhu had had his revenge.

It was a few days before the Irilas were able to gather enough courage to visit the swami in a deputation, headed by Kantha, the owner of the banana plantation. They prostrated themselves at his feet, begged forgiveness, and asked him to beseech his familiar, the tiger, to spare their lives; or, if he happened to be the tiger himself, to kill no more of them. The Irilas suggested that, if human prey were really essential to his diet, there were plenty of Badaga settlements a mile or two higher up the ghat road, or Karumba settlements a bare ten miles to the north, near the border of Mysore State, to appease his appetite.

The swami appeared to listen to the deputation patiently, thought for a while, and then said that, if the Irilas agreed to pay him a trivial stipend of a hundred rupees each month, in addition to a quota of rice, ghee, vegetables and, of course, the bananas, cream and sugar as before, he would consider visiting the Badagas or the Karumbas, or both, in preference to the Irilas.

That settled it, they agreed. Had he not said he would consider transferring his attentions? Without doubt the swami himself was the tiger!

But where, and how were they to raise the hundred rupees, the ghee, the rice and the vegetables? They muttered and whispered among themselves, then attempted to bargain for fifty rupees. The swami was adamant. Silently they stole away; a hundred rupees was too high a price to pay.

Four days later the tiger took another Irila. He was snatched a furlong from his hut; and that, too, at noon. Also, very significantly, he had been one of the members of the deputation.

* * *

I admit the preamble to this story has been long; but I wanted to give a clear picture of the simple, superstitious minds of the aboriginal folk who dwell in the Indian jungles, and I thought the best way to do so was to tell the whole story, just as it was told to me beside the camp-fire by the bank of the stream, a mile below where it skirted the Valaithothu grove.

My informant was a short and grizzled old Irila named Boora, who had served me previously when I had visited the area on hunting excursions.

The aborigine is usually very reticent to strangers and one has to acquire the knack of getting on with them before they abandon their natural shyness and launch into unrestrained conversation. That knack lies mainly in putting the jungle-man completely at his ease, behaving so naturally that he feels you are quite at home with him. He then reciprocates by becoming at ease himself when, if you are a good talker, you can induce him to tell you many interesting things about the jungle, its animals, his people, the aborigines and their habits, folklore and superstitions.

I well remember that night in the jungle, under a far-spreading mango tree, when Boora told his strange tale. We had heard the twang and crash of giant bamboo stems being torn down by wild-elephants. This, and of course the presence of the man-eater, had caused us to light a big fire, and to keep it burning, fed by a large pile of dead logs we had gathered before sunset. It was cold—very cold, despite the leaping flames. We were huddled side by side, Boora feeding the flames every now and then, while I stared steadfastly at the crackling logs on their mound of glowing embers.

A close friend of mine, Martin Satsell, living at Ootacamund, had casually written to me about the tiger and had suggested I visit him first, spend a day at Ooty, and then walk down the twelve miles of ghat road to the banana-grove to find out about things for myself.

I had done just that and had arrived at the banana grove at about four in the afternoon. Boora, who lived nearby, expressed pleasure at seeing me again after quite a while, and we had decided to spend the night by the fireside under one of the large wild-mango trees on the banks of the little river.

Looking furtively over his shoulder into the inky darkness of the jungle beyond the faint circle of light from the leaping flames of our fire, with bated breath and as if afraid of being overheard, Boora told me the whole story, just as I have told it to you. Then, continuing in Tamil, he ended his narrative thus:

'Perhaps I should not have told you all this, *dorai*. It is said that the swami-tiger and the spirit that protects him, as it does every man-eater in a jungle, has ears everywhere and can overhear the faintest of whispers, if not read a man's very thoughts before they are spoken. In that case, the holy one will be angry indeed and my life will surely be forfeit.'

As if in eerie response to his words, at that moment a tiger roared loudly from across the stream and within a distance that I judged to be only a furlong: 'Oo-on-gh! a-oongh! aungh-ha! aungh-ha! ugh! ugh!'

Boora stiffened beside me and then trembled like an aspen. The words stuck in his throat as he croaked, '*Dorai*, it is he! It is the devil tiger! The swami—the holy one—has overheard me. I am doomed!'

There was a nasty tingling, prickling sensation at the nape of my neck as the short hairs there persisted in standing erect, and I felt the sweat break out all over my body. But putting on a show of nonchalance, I said unconcernedly: 'Nonsense, Boora, your words are but idle superstition. That was just an ordinary tiger, of flesh and blood. Maybe it was even the man-eater. But certainly no reincarnated or transformed human-being, spirit or devil.'

Involuntarily, I laid my loaded .405 across my knees, and picking up my three-cell torch, I shone it into the darkness in the direction of the calls. Paled by the glow of the fire, the little circle of light seemed pitiful in all that gloom.

We waited in silence. Some time later, the tiger called again. This time from the same bank of the stream as ourselves and from behind us. He had waded across the shallow water and had come around in a half-circle.

I threw another chunk of wood into the flames, wondering if our fuel would last till morning. We had certainly laid in a good stock, but with the tiger's near presence we would have to use more firewood than we had anticipated.

There was silence after that second call. Deep, prolonged and uncanny. We began to long for the tiger to roar again. At least, if he did so we would know his approximate whereabouts. This way, with no sound whatsoever, the beast might be a mile away or behind the nearest bush, a few yards distant.

Stealthy rustlings, inevitable in every forest at night, came to us from beyond the circle of light cast by the feeble flames of our fire, and we conjured up visions of a dreadful man-eater stalking us, creeping ever forward, ever nearer, inch by inch, till he came within striking distance. Then would follow a short spell of agony and afterwards oblivion!

We waited for the sudden but expected charge. Stricken by fear, Boora kept up an almost unintelligible mumble. Now and then I heard the words 'revenge' and 'neither of us will live to see the sun rise'.

This got on my nerves, frayed as they were by the story I had just heard from my companion, closely followed by the tiger's nearby roars, and I almost shouted at the Irila to shut his mouth. But this he evidently could not do, and his incoherent mutterings continued.

A sambar stag bellowed on the stream-bed, 'Whee-onk! oo-onk! dhank! dhank!', a few seconds intervening between each call. He was clearly very agitated and had either seen or smelt the tiger close by. Minutes later a bull-elephant gave vent to his trumpeting scream of fear, tinged with hate: Tr-i-a-a-ank! Tr-i-a-a-ank!', and the young elephants in the herd screeched, ' 'Quink! Quink! Quink!', like piglets, as each sought shelter under its mother's ponderous bulk.

The tiger was walking around us in a circle. He knew there were men within a few yards, but the light of our fire deterred him from coming closer to investigate.

Then there was a continuing scream of terror, strident and loud, from the direction where the sambar had last called. With all his vigilance, the stag had not been watchful enough. The tiger was killing him!

In the deep silence that fell all around us with the end of the sambar's dying screams, the herd of elephants, tightly bunched together with their calves in the centre for protection against a sudden charge by the prowling marauder, had evidently melted silently away. There followed a nerve-wracking period of tension till midnight, when the rising moon dispersed some of the dense blackness under the huge trees around us that clothed both banks of the stream. Splotches and splashes of silvery moonlight fell upon the ground and the surrounding foliage as it struggled through the close canopy of branches overhead. The darkness was not so intense now and we could at least see a little. Nor was there any evidence of the tiger. Boora and I became somewhat more composed.

But he was in no condition to be entrusted with the responsibility of keeping watch by himself, and so I spent a sleepless night, the murmur of the water, burbling and bubbling and rippling over the boulders on the stream-bed lulling me towards a slumber I struggled hard to combat.

The false dawn arrived with the still-lingering moonlight, deceiving a solitary jungle-cock into thinking that daylight was at hand. 'Wheew! kuck-kya-kya-kuckm!' he crowed once and then fell silent, as if ashamed of his mistake.

Then came the real dawn, heralded now by his companions in all directions, and he joined lustily in their calls. Indescribably beautiful were those first few minutes of early light. 'Mia-aow-iao! aio-aow-iao!' screamed the peacocks in the valley beyond. 'Kuck-khera-wack! kuck-khera-wack!' wrangled the quarrelsome little spur-fowl in the undergrowth bordering the water. A black-and-white dayal bird, sitting on a tree-top, sang loud and joyously to his mate, huddled on her eggs in a hollow of the trunk; and the bulbuls chirruped and flapped in hundreds while gormandizing on the clusters of small black berries that showed between the rough leaves of lantana bushes.

The eastern sky, that had been growing lighter, put on a sea-green tint. This dissolved into layers of many colours as the low clouds on the horizon reflected and refracted the rays of the rising sun. Bewitching tones of violet condensed into a suffusion of deep indigo, while below it a wider band of ethereal blue separated a belt of turquoise green. Just above the serrated line of the jungle trees, some of them partly hidden by the wisps of swirling mist that rose from the damp floor of the jungle, shades of canary yellow melted to a flaming orange and then merged into a band of angry blood red.

This play of many colours lasted only briefly, to fade as suddenly as it had formed, when above the tree-tops arose a segment of blazing light—the sun!

Another day was born.

I was sleepy, and I was annoyed at having thus carelessly put myself in a position where I could do nothing. But who would have expected to hear so strange a tale? And who would have thought the lone tiger would make himself heard, if not felt, so soon? Curiously, I was not angry with the tiger; but I definitely was angry with myself, and with the sadhu who had so frightened these simple jungle-folk, and I determined as soon as possible to confront the hermit in his own hut.

Brewing a little tea, which I drank while munching the sandwiches my friend Satsell had given me in Ootacamund, I told Boora very clearly that he was a fool to listen to such tales, and that I would soon show what sort of rogue the sadhu was. To that Boora replied in simple words: 'Till this moment, dorai, there was just one of us doomed to be killed by the holy tiger. That was me. Now, after your words, there are two, dorai. The second is you!'

That made me even angrier.

It was just seven in the morning when I stood before the closed door of the fakir's grass hut, Boora shamefacedly behind me, and called loud and aggressively: 'Open, O holy tiger; that is, if you are awake yet after your night-long ramblings.'

As if he had been expecting my visit, the sadhu flung open the door and stood before me, dressed in a flowing robe of saffron hue, his only ornament a heavy necklace of amber beads, his long and oval face enhanced by his flowing black beard and fierce moustache.

His expression displayed intelligence and determination.

In even, pleasant tones the fakir answered: 'As a matter of fact, I have just awakened, my friend. I expected the white man would visit me after hearing the tale told to him by that jackal who now slinks behind you. I saw you both by the fireside on the river-bank last night and was tempted to eat one of you, but the sambar that called did not get away soon enough, and I ate him instead.'

To say the least, I was thunderstruck! The man spoke as if he had really played a part in the previous night's drama. I knew, of course, there was a trick somewhere. No man could turn into an animal and then turn back again, although I had heard such things many times from the jungle aborigines; I had also read of similar occurrences in tales from Africa. Here seemed to be the real thing.

I decided to act tough and take the sadhu down a peg or two.

'Listen, O hypocrite, rogue and liar, dressed in holy robes,' I began hotly. 'Your silly stories may frighten these simple Irilas, but on me they have no such effect. You are a man of the meanest nature, having repaid the hospitality of these humble, honest folk with lying threats and greedy blackmail. Let me tell you this, O trickster-whose-game-is-up. The tiger that called last night is at least a gentleman; you are nothing but a swine!'

He laughed. 'Why is the hunter so angry? Is it because he knows he is helpless?'

Then his mood changed and the handsome features contorted to a scowl. 'Mind your own business, white man, and interfere not in what does not concern you and you do not understand. For if you do, you will pay with your life. I have spoken!'

An almost uncontrollable urge came over me to smite that rascal with no further delay. I remembered in time that he was regarded as a pious man and such an action would definitely land me in trouble. So instead I said, 'Take it for granted that I shall interfere, O teller of lies; and when you become a tiger, see to it that you do not show yourself before the sights of my rifle. For if you do, I promise that you will indeed be a very dead tiger.'

'Or you a very dead hunter,' he rejoined quickly.

Then the fakir slammed the door of the hut in my face. I turned on my heels and strode away.

At my side, Boora resignedly pointed out several fresh sets of tiger pug-marks. They were hardly half-a-furlong from the hut I had just left.

I led the way back to the wild-mango tree where we had spent the previous night, opened my bedding, which I had not used, and decided to snatch a few hours' sleep. But it was a restless, unrefreshing slumber, disturbed by dreams of a tiger with a sadhu's head.

Shortly after noon I awoke, ate some tinned sardines with the remainder of the stale chappatties I had brought from Ootacamund, and while the hot water for tea was still brewing, sat with my back against the tree to consider the situation.

Although I did not for a moment credit the strange tales that were being told about the sadhu, I had to admit I was confronted by a very curious sequence of events. The sadhu had said he had seen Boora and myself the previous night while he was in the form of a tiger, and would have killed one of us had not the sambar stag got in his way. And a sambar had indeed been killed the night before. I had heard the stag's death-scream with my own ears. How did the old devil know all this? Obviously, he could not have turned into a tiger. No man could. There were only three possible explanations. The first, someone had told him. The second, he had been watching and listening and had seen and heard for himself. Thirdly—and I abandoned the idea as soon as it came to my mind—the tiger had been his informant. That, too, was ridiculous.

Nobody could have told the hermit, for nobody would dare to be afoot in the jungle at night with a man-eating tiger roaring nearby, not to mention the wild elephants. Boora had heard; but Boora had been with me all night.

That left only one solution. The sadhu had been watching us and had heard the dying cry of the sambar. If that was really so, I had to admit he was a brave man. To be hiding in the jungle unarmed, spying on us, with the man-eater circling the whole scene, did indeed call for nerves of no mean order.

It did not require much thinking on my part to come to what I considered the correct conclusion. It was the only common-sense explanation possible. The sadhu was an opportunist, pure and simple; albeit a decidedly clever and courageous one. No doubt, coming to learn from the Irilas that his advent to the locality had been followed by the arrival of a particularly bold tiger, he had very astutely decided to link the two events by creating conditions and circumstances that would lead to the opinion that he and the carnivore were on close and familiar terms, thus enhancing his reputation in the area and providing him with free board and lodging for the remainder of his life.

Fate had then played into his hands by bringing about events that brought him into even closer relation with the lone male tiger that had come to stay. The sadhu had either been injured by coincidence when Kantha, the owner of the banana tope, had wounded the tiger, or perhaps had not wounded it—or the fakir had somehow come to learn of the shot that had been fired at the tiger, and had deliberately inflicted some slight injury on his own arm, sufficient to draw blood from himself and give birth to the story that he and the tiger were one, and that he could change himself from human to animal form and back again at will. Such a reputation would ensure life-long fear, respect and a status beyond all chance of question or reproach, together of course with all the inevitable material benefits that resulted from such distinction.

Then the tiger had become a man-eater. The sadhu had been clever enough to exploit even this development by levying the payment of a hundred rupees, together with various commodities, as the price of leaving the Irilas in peace. No doubt he anticipated that, if and when the tiger killed more Irilas, such happenings would simply enable him to raise his demands, both in cash and in kind. His future financial prospects at Valaithothu were bright indeed!

It was easy to understand from all this that my arrival on the scene represented a serious spoke in the wheel of good fortune that had thus far favoured him and looked like favouring him as long as the man-eater remained in the vicinity. For if I succeeded in killing the man-eater while he remained alive, this would once and for all explode the belief held by the Irilas that he and the carnivore were one and the same. Not only that, but the sadhu could never again hope to create such a train of favourable circumstances here or elsewhere. This would be a serious blow to his reputation, and would ultimately affect both his purse and stomach. People would pay him nothing, nor give him food. Doubtlessly, the banana-grower would lose all fear of him and turn him out of the grass hut he was living in. And where would he ever get such delicious bananas again, spread so thickly with fresh cream and plenty of sugar?

Indeed, everything pointed to the fact that at all costs I would have to be got rid of in some manner. The fakir had no doubt learned from my words that very morning that his threats would not frighten me away. Therefore I could expect that he would resort to more direct action.

If only something could be made to happen to me, what a boon it would be for his reputation! It would soar sky-high, indeed! People all over the district would wonder at the powers of the terrible sadhu, who could turn himself into a man-eating tiger at will, and by whose magic even a white hunter had been eliminated.

I realized that I was faced with two implacable enemies: a man-eating tiger and an unscrupulous human, who would stop at nothing. And I knew that, of the two, the human was by far the more dangerous—because he could think. He could rationalise and anticipate my actions, and the Irilas would unwittingly help him by keeping him informed of my movements. The man-eater, on the other hand, had only his instinct and jungle skill to rely upon. He did not have the capacity to reason.

I also knew that, whatever plans might be devised to trap the wily tiger, and incidentally the even more wily sadhu, they would have to be devised and executed entirely by myself, unaided and in secret. I could not afford to take any of the Irilas into my confidence, not even my ally Boora; for although I knew they would be loyal to me, their superstitious fear of the fakir, combined with the inherent long-tongue possessed by all simple village and jungle folk, would cause them to be garrulous at the least opportunity, revealing my doings to their friends. The fakir would, in fact, soon come to learn my plans.

After some thought, I felt it would be the best to begin operations against the man-eater first, regardless of the fakir and what he might do; because, if I succeeded in killing the tiger, the fakir's reputation would automatically fall apart.

Finishing the tea I had brewed, with Boora following, I led the way to the nearest patti to buy a couple of buffalo-heifers to tie up as baits for the tiger that night.

That was when I really began to understand the malign influence that was being exerted by the hypocrite in the saffron robe. Doubtless he had already anticipated my move and had taken steps, by preying upon the superstitious fears of the Irilas, to thwart me. For the herdsmen flatly refused to lend, hire or sell a single animal.

I tried hard to explain to them that, after all, my efforts to kill the man-eater were in their own interests and for their safety. They avoided giving me a direct reply to that point when I mentioned it, but countered vaguely by saying that if they rendered any assistance to anybody engaged in attempting to kill the human-in-tiger-form, as they called him, swift vengeance would fall upon them and their herds.

Seeing that further argument was useless, we visited three more patties in the area, but the herdsmen in each case repeated the same excuse when declining to help in any way. A grey-beard at the last of the patties, seeing me turn away in anger, called to his companion and ordered him inside the hut to attend to some trivial task. As soon as the man was out of sight and hearing, the ancient sidled up to me quickly and said, '*Dorai*, if you really want live-baits, go to Masinigudi hamlet, five miles along the road, and get them from there, for nobody here dare part with a single animal to help you. although we would like to do so.'

I understood the significance of his words and pretended to leave in a great rage, just as his companion returned. The only possible thing to do was to follow the advice the old man had gone to the trouble of giving me so secretly.

But it was too late that day to go to Masinigudi and return before darkness set in. I would have to spend another night in idleness.

I had come at a bad time altogether for tiger-hunting, due to the fact that the moon would rise only in the early hours of the morning. To sit on a tree in the darkness without a bait was futile. So, till such time as I could procure some from Masinigudi, there seemed no alternative but to spend another night by the fireside beneath the old mango tree.

This time, Boora and I started early and by dusk had laid in a very substantial stock of firewood, in the form of dead branches and fallen logs, with which to replenish our camp-fire and ensure that the flames were kept up until morning.

We lit the fire when the shades of twilight deepened, for the man-eater might already be afoot, and in that uncertain light he could stalk us by creeping under cover of the dark shadows cast by the towering fronds of bamboos, and the tall muthee, jumlum and wild-mango and tamarind trees that bordered the stream, to within striking distance, of where we sat. After that he would pounce, and we knew we could not see him coming.

The calls of the roosting birds on this occasion entirely failed to bring the accustomed content to my listening ears, for the simple reason that I had not been listening, my attention being divided between gathering firewood for the hours of darkness and a sense of mortification that the sadhu had outwitted me by compelling me to spend another night of idleness without even a single live-bait to sit over.

'Chuck! Chuck! Chuck! Chuck! Chuck-ooooo!' The nightjar, invisible now in the darkness, was calling from the further bank of the stream, and many of his kind soon joined in jerky chorus from all directions. It seemed as if a whole company of schoolboys had been let loose in the jungle and were deliberately throwing marbles upon a glass floor, each marble bouncing four times before slithering to rest.

The chorus was interrupted by the harsh scream of a night-heron in the distance, followed far closer by the almost-human cry of the 'herd-boy' bird. 'Oooo! Oooo! Oooo!' it called, imitating to perfection the cries of the herdsmen as they drive their cattle to shelter while the sun has yet a long way to go before it sets behind the tall spurs of the Nilgiri Mountains to the west.

The time was approaching eight o'clock when a hushed stillness suddenly fell over the jungle. Even the crickets seemed to stop their pulsating chorus of 'Sizz! Sizz! Sizz!' There was not a sound. Abysmal silence reigned over everything. The forest appeared empty of life, except for Boora and me.

This unaccountable lull acted as a prelude to an event that was as unnerving and unexpected as any I have ever experienced. For suddenly the tiger moaned three times in the darkness, somewhere by the stream to the south of us. 'Oo-on-gh! O-o-o-o-ongh! A-oongh!'

Silence again, but for barely a minute. Then from the west of us, in the direction in which the road from Ootacamund led away to Masinigudi, came an unearthly cacophony of sound, not loud, but clear: 'Aaa-ha-ha-ha-ha! Aaa-ha-ha-ha-ha! Aaaaa-hah-hah-hah!'

Boora, beside me, literally shrank with terror, while the eerie noise caused the tingling sensation of fright to race across my scalp.

Then the spell was broken. Grabbing my torch and rifle, I leapt to my feet and raced towards that unearthly din. Gone were all thoughts of the supernatural. Commonsense had asserted itself as I realized that the sadhu had blundered badly by giving himself away. For the tiger that had just been calling from across the river could not have covered the distance to the road direction in so short a time.

The fakir was trying to frighten me. No sooner had the tiger called than he had given vent to that raucous laughter to make us think the carnivore had already changed to human form. He had given himself away, and I intended to catch him red-handed.

Following the wavering beam of my torch, I dashed along the narrow footpath that soon led to the roadway. There I turned to my left, and pelted towards the banana-tope and incidentally the fakir's hut, only a short distance away, hoping to overtake the rascal before he could reach it.

But evidently the fakir could run far faster than myself, for I eventually arrived at the hut, which was in utter darkness, without a glimpse of him. I tried to wrench the closed door open. It resisted my efforts and I banged with the flat of my hand against the surface of closely-spliced bamboo.

After an irritating delay I saw through the gaps in the bamboo the flicker of a lamp within. The light approached and halted, and there passed a few moments during which, presumably, the fastenings were being undone. Then the door flew open, and before me stood the hermit. This time no saffron robe adorned him. He was naked but for a brief loin-cloth and the string of large beads that dangled from his neck.

I shone the beam of my torch directly in his face.

Never for a moment did he close his eyes to avoid the light, but his lips curled in a derisive smile as he rasped: 'The white man can still run well for his age, but did he think it possible to catch the tiger's spirit when it laughed at him?'

Exerting a tremendous effort to control myself, I replied, 'You run still faster for an even older man. But beware, O trickster, the man-eater does not one day decide to end this farce by eating you.'

Before he could reply, I swung around to walk away and bumped into old Boora who, unknown to me and afraid of being left alone, had followed hot-foot in my steps. Without a word and in deep chagrin. I walked back to our camp-fire, my attendant close behind me. No thoughts of the man-eater entered my mind; only vexation at the ignominious way in which the sadhu had again made a monkey out of me.

We spent another sleepless night, during which nothing happened, and in spite of our tiredness set out for Masinigudi just after sunrise with the avowed intention of purchasing three live-baits, which I intended to tie up for the tiger.

With some considerable difficulty, and of course a great deal of bargaining and bartering, I managed to procure two young buffaloes and a half-grown brown bull, which we led back to our camp-site with the minimum of delay. It was then one o'clock in the afternoon and I felt very, very tired after the two sleepless nights we had already been through. To keep awake for another night would be impossible and so, indifferent to the man-eater's possible proximity, we tethered our three animals at the base of the old mango tree and fell asleep amongst them.

I awoke at exactly four o'clock. The short three hours of slumber had nevertheless dispelled the sense of numbed exhaustion that had been creeping over me, and I determined, with only two and a half hours of daylight left in which to make some sort of preparations, to try to meet the tiger on equal terms that night.

Now I want you to follow me carefully for a little while in order to understand clearly the lie of the land, for that has a very important bearing on what transpired subsequently.

The main range of the Nilgiri Mountains lay to the south of us, but a spur, in the form of a lower line of hills, jutted out to the west. At the base of this spur, and parallel to it, was the road from Ootacamund, leading from south to north, towards Masinigudi. Then came a strip of scrub jungle, varying in width from a few hundred yards in some places to a mile, separating the road from the stream which zig-zagged along from the foot of the main range of the Nilgiris to its confluence with the Moyar River over ten miles away to the north-east. All along and beyond the river was dense jungle.

The Valaithothu plantation was at the base of the main Nilgiri range and adjacent to the river, and as I have said the man-eater appeared to have his home somewhere in this vicinity. That was common talk, confirmed by the fact that on the two previous nights his first roars had come from this direction.

Therefore, if I tied my baits in such a fashion that one of them was within sight of the road shortly after it had passed the banana-grove, the second within sight of the river bank and to the east of the road and my first bait, and the third at the base of the low spur of hills that ran parallel to the road on the west, from whatever direction the tiger approached he would be bound to blunder into one or the other.

The only mistake in this plan was that, as matters stood just then, there was insufficient time for me to tie out more than one bait that night and to build more than a single machan before the sun set.

Consequently, I would without delay have to make a selection from one of these three positions and build a hasty machan nearby. The other two animals I would ask Boora to take back with him and tether outside his home that night, till we could make use of them the following day.

Now came the question: which was the most promising place to tie the baits? For the last two nights the tiger had called by the riverside. As a hunter he was not likely to use the same approach for three consecutive nights. This time he would in all probability come either along the road, which would also bring him near the sadhu's hut, or he would approach from further west, in the direction of the parallel spur of hills.

I remembered that, the morning of the previous day, Boora and I had seen the tiger's pugs about half a furlong from the fakir's hut. The previous night he had not come that way.

All the above facts, taken together, seemed to indicate that in every likelihood the tiger would approach from the west that night. Of course, there was absolutely no certainty of this for he might just as well come along the road near the sadhu's hut. It was purely a gambler's chance, backed by a modicum of reason, but my best bet seemed to be to tie to the west, at the foot of the spur of hills.

Making my mind up quickly, I joined Boora in driving all three baits to his hut, a mud-and-wattle affair situated in a small hamlet of half-a-dozen similar huts within two furlongs of the banana plantation. Here we left the two buffalo-heifers and, selecting the brown bull-calf as the animal most likely to entice the tiger, drove it between us to the base of the foothills in the west which, at that spot, was a little less than a mile distant.

We reached our objective, a nullah that I knew ran down the hillside and turned eastwards to join the river we had just left. Clambering down to the bed of this nullah, which was dry and nowhere more than ten feet deep, we found many pug-marks of the lone tiger clearly visible in the soft sand. But not one of them was less than four to five days old. These tracks led in both directions, proving that the nullah was a favourite route, followed by the man-eater on his hunting-trips in search of animal victims, and these signs made me feel more optimistic about him choosing the nullah that night.

On the top of one of the banks of this ravine grew what is familiarly known as an 'Indian-oak' tree. It has large leaves, growing in pairs, with clusters of flat bean-like seeds, and is a favourable tree for tying a machan, as it has no thorns but does have a rough bark that enables the ropes that hold the machan to grip firmly.

We decided to tie the bull-calf about fifteen feet from the base of this tree, but near the edge of the ravine, so as to be clearly visible to any tiger either walking down the nullah-bed, ten feet below, or stalking along the top.

Having, as I have already said, large leaves, there was no necessity for much elaborate camouflaging for the machan which was readily—and very luckily—supplied by a convenient fork about ten feet off the ground, where the three main branches of the tree bifurcated from the trunk. A little filling up of the hollow where these branches separated with fresh leaves that would not rustle should I move, together with a few fronds and tendrils of the large leaves plucked from the higher boughs of the tree and woven in between the three jutting limbs, formed a hurried although far-from-perfect machan.

But by this time it was gone half-past five and we could delay no longer. For one thing, Boora had to make his way back to the settlement unarmed and alone. Secondly, even at that moment, should he be lying up anywhere along the spur of hills to the west, from his elevated position the tiger would be able to see us clearly moving about near the tree. If he did, it might have two quite opposite effects. Should he be daring enough, as he certainly appeared to be, he would make straight for the spot to see if it was possible for him to find a human tit-bit for dinner. On the other hand, should he have the inherent cowardice and deep suspicion common to practically every man-eater, he would give the place a very wide berth and not come anywhere near it that night or for several nights thereafter.

Boora tied the brown bull to a convenient stump within thirty feet from where I was crouched, performed the usual farewell salutation, asking its pardon for the part he was playing in bringing about its death by prostrating himself before the animal and touching the earth thrice with his forehead, and then hastily went his way.

The sun had long disappeared behind the intervening hills to the west, and the shades of evening began to fall by six o'clock. I sat facing the hills as being the most likely direction from which the tiger would appear. Behind me, from the bed of the river, an elephant trumpeted mildly, no doubt enjoying his early evening bath. Unafraid of this huge but to them quite harmless companion, the jungle-fowl, spur-fowl and pea-fowl kept up a recurrent chorus of calls from the aisles of bamboo that grew in profusion on the further bank of the stream. Up above me on the hillside immediately in front, a troupe of black-faced, long-tailed langur monkeys romped and gambolled in the sheer joy of living.

'Whoomp! Whoomp! Whoomp!' The slopes echoed and re-echoed to the joyous cry of the males, interspersed occasionally and faintly, now and then, with a sharper, more timid, 'Cheek! Cheeek! Cheeeek!' from the females and young. How I envied those monkeys at that moment! To them life was all play, with no responsibilities. In faith they capered from tree-top to tree-top, regardless of their ever-watchful enemies, a possible panther or tiger lurking below, relying on the sharp eyes and alert senses of their sentinel, the solitary langur-watchman, posted by them at the summit of the tallest tree, ever watchful for the slightest suspicious movement on the floor of the jungle below—the deeper shadow that might betray a furtive prowler—or the faintest crackle of bramble or brush that would betoken the stealthy footsteps of a spotted or striped killer.

And it was that very langur-watchman who gave the first alarm before the sun had even set. Floating across the cooling air came his harsh, harrowing, warning call: 'Ha-aah! Harr! Harr! . . . Ha-aah! Harr! Harr!'

That monkey-watchman did a magnificent job, for he not only alerted his troupe of monkeys, who ceased their romping at once, but he sent the spotted-deer, grazing in the valley, helter skelter for shelter with sharp cries of 'Aiow! Aiow! Aiow!'

And he alerted me. I knew the man-eater was coming, and I hugged myself with pleasure at having reasoned out his line of approach correctly.

Or so I thought!

For the tiger confirmed his presence. 'A-oongh! Aungh-ha! Aungh-ha!' he moaned from the hillside facing me.

But he did not turn up, and after fifteen minutes of hysterical harr-harring, the langur-watchman fell silent and the spotted-deer ceased their sharp cries.

Darkness was falling rapidly, and while sitting perfectly still I strained my eyes by staring into the gathering gloom, expecting at any moment to detect the form of the tiger slinking down the nullah.

Within a few minutes an animal did hasten down the ravine with urgent, rolling gait, but it was only a sloth-bear, so intent upon reaching the fallen, ripened boram fruit in the scrub-jungle to the east that he never for a moment thought about the tiger, and certainly did not notice me perched on the Indian-oak tree.

Then night fell. The air in the whole valley where I sat, heated by the day-long sunshine, began to rise, and the cold air from the mountains rushed down to fill the vacuum. It grew bitterly cold, I became numb and uncomfortable, and I had difficulty in keeping my teeth from chattering.

Suddenly, I sensed a faint movement behind me. Yes, there it was again! Distinctly I felt the touch of tiny, fairy-like fingers above my elbow and then on my neck. The thing, whatever it was, was perched on my shoulder!

Quickly I jerked my head around to see what sort of elfin visitor had thought fit to touch me. A pair of large, soulful eyes gazed into mine from a tiny head hardly two inches across. A miniature human-like body, not much more than six inches tall but with disproportionately long arms and legs, was sitting comfortably on my shoulder.

But my new companion turned out to be no fairy, come to warn me of impending danger; he was only a 'slender lorris', a delicate monkey-like animal, entirely nocturnal in its habits, that lives in the jungles and waste-lands of Southern India.

Incidentally, these unfortunate little creatures are much persecuted by certain ignorant aborigines, who believe their large and lustreful eyes have wonderful properties to cure venereal complaints. The poor beasts are caught and their eyes are then gouged out while they are still alive. Afterwards, they are killed and eaten as a further part of the remedy. But to be really efficacious, the jungle-men believe the eyes must be pulled out while the animal is still living. Truly, some of the unfortunate creatures of the forest suffer greatly at the hands of men!

The lorris took fright when I moved my head. Quickly it clambered up one of the branches at my side and disappeared. Minutes later, when the little lorris found that it had not succeeded in escaping from danger, it decided to voice its screeching alarm call. 'Keee-eee-eee! Keee-eee-eee!' came the sound from the tree-top above. The tiny monkey was really worried: he had climbed up the tree only to discover he could not escape without climbing down and passing over me again.

The hours dragged slowly by in that extraordinarily silent night. Except for the soughing of the sharp gusts of chill wind blowing down from the mountains, the forest appeared to be bereft of most forms of life. Periodically, a wood-cricket chirruped from somewhere in the ravine, and once a bull-bison whistled a furlong or more to my rear, as a puff of breeze carried my human scent to his keen nostrils.

The sky was cloudless, spangled with a million stars, a sight to experience and remember, defying description by pen or tongue. The combined light of the heavenly bodies dispelled the darkness and shed a faint glow over the vista of jungle and hillside around and above me.

At long last the silence was shattered by a discordant medley of sounds from the lower ground leading to the valley behind, through which the river meandered: 'Oooo-ooh! Wooo-ooh! Oo-where? Oo-where? Here! Heere! Hee-yah! Heee-yah! Yah! Yah!'

It was the cry of a jackal pack, furtively searching for such food as they could find. The leader voiced his weird call, and the other jackals echoed it as they trailed along behind.

Then, in mid-cry, the pack broke off short at the very climax of one of their choruses. Not a whimper, not a sound followed. Why was that?—I began to wonder.

The reason came in another five seconds, the sound floating from behind me from the very direction where, but a moment before, the jackals had been in full cry.

'Ugh-ha! Oo-ooongh! Aa-oongh!'

The tiger had come down the hillside and passed my position on the broad-leafed, Indian-oak tree. From his roars, he was now probably on or near the road, and close to the banana-grove . . . And near the sadhu's hut!

I regretted that lack of time had prevented me from tying one of the buffalo-calves in the vicinity of the road, as I had originally intended. That was when a perfectly insane notion entered my head. Why it did so bears no justification whatever. It just did.

Now that the tiger was calling near the fakir's hut, I felt impelled to get down from my tree and hurry there as quickly as I could. Maybe I was beginning to half-believe old Boora's weird story and wanted to see for myself whether the sadhu was at home or not.

The tiger was still calling as I scrambled down the tree and jog-trotted towards the hut. It was less than a mile away, but so long as I could hear the tiger's roars I knew I was safe from a sudden ambush. A segment of moon had arisen about an hour earlier, and by its light, although poor, I was able to retrace my steps along the footpath I had followed that evening on my way to the nullah.

I reached the hut from the west as the man-eater roared somewhere near the roadway to the east. The hut was in darkness.

Was the fakir inside? Creeping up to it, I placed my ear to the wall of plaited bamboo, endeavouring to catch some sound from within. But nothing could I hear. No breathing, as of a man in deep slumber. The hut appeared to be empty.

That was when the tiger roared once more. He was very near now. In all probability walking along the roadway, hidden by the closely-growing banana trees, less than a furlong to the east.

For the second time that night, an idea came to me on the spur of the moment, and without waiting to consider its merits I acted.

Filling my lungs with air, and cupping my hands over my mouth, I answered the tiger, imitating his call in a manner I had practised under the able tuition of my old friend and instructor, Byra the Poojaree.

'A-a-aungh! A-a-a-aungh! O-o-o-ongh!' I called deeply.

Then I remained silent, crouching behind the corner of the bamboo hut, rifle half-lifted to my shoulder, with the ball of my left thumb lightly touching the button-switch controlling the three-cell electric torch clamped to the barrel of my trusty .405. The tiger never roared again. He just walked straight towards the sound he had heard. I knew he was coming all right.

A deep silence fell all around. It was an oppressive silence that could almost be felt, and a strange unaccountable sense of foreboding began to creep over me. Why this should happen I could not tell, for I was safe enough in my position around the corner of the swami's hut. The moonlight shone down palely upon the jungle around me, just strong enough to make things visible without making them clear.

It might have been a minute or two later when the unexpected happened. A human figure broke cover from the cluster of banana trees, in exactly the direction from which the tiger had been calling, and came running towards me. It was a man, and he at first appeared to be stark-naked until, upon closer approach, I could see that he was wearing a dark loincloth.

Then I realized that the man was not running towards me, but towards the hut, for I would be quite invisible to him from where I was hiding, around the corner. And the man was the fakir himself!

My breath caught as I grasped the significance of what was happening before my very eyes. Hardly a few minutes earlier I had heard a tiger roar by the roadside. Now here was the fakir, coming from the very spot where the tiger had been calling!

It was true, then! I was witnessing a scene as bizarre and strange as any man could have seen with his own eyes. But the very next moment, the mystery was rudely shattered and the black magic gone!

For, with a series of short roars, the tiger charged from behind the fleeing figure, and in two bounds caught up with him. The unfortunate man's shriek sounds in my ears to this day: 'Ai-yo-ooo-ooo-ooo!' he screamed, and that was all.

As the man-eater leapt upon him, tiger and man disappeared from sight in the long grass in the pale moonlight. The unexpectedness of everything had rooted me to the spot. Now the spell was broken, and with a loud yell I ran forward as fast as I could run. Perhaps I might yet be in time to save the swami; perhaps my shout might frighten the tiger away.

I had hardly forty yards to go. I pressed the button of my torch as I ran, but it was only from a distance of fifteen yards that the beam fell upon the tiger's glaring eyes as he crouched over his prostrate victim in the long grass.

Without thinking, I stopped abruptly in my tracks and fired between those terrible eyes. The tiger half-rose, and I fired again and yet again into that murderous head. He collapsed, and a great quivering, as if from a terrible ague, shook his whole frame.

Only then did I realize that the man-eater had fallen on top of his victim and practically hid him from view. More terrible still, any of my three high-velocity bullets, fired at that short range, might have gone through the tiger, or even missed him, and entered the prostrate fakir below.

I went up to the tiger, still twitching spasmodically, sat on my haunches and fired a fourth shot from a distance of two yards into the back of his skull, already shattered by my earlier bullets. It was another minute before the twitching ceased. Then, with some difficulty, I pulled the dead tiger off the man on whom he was lying.

The fakir was covered with blood and a cold fear overcame me. I was sure I had shot him. He was lying on his face on the ground, and there were no signs of breathing or movement of any sort. Quickly pulling my torch from the clamps that held it to the rifle-barrel, I began to search for the bullet hole, or possibly holes, that I felt certain I would find. I looked carefully enough, although my hands trembled violently. There were none. Then I turned the swami over. Possibly I would find what I was looking for on the other side.

That was when the fakir opened his eyes. They glittered in the torchlight with even greater hatred than the man-eater's. His lips moved slightly and I stooped to catch his words. They came faintly, haltingly, but I heard them all the same.

'You win, white man; but the stars were against me this night. The tiger has taken my life; but so shall you die, too, and I shall await your death at another tiger's jaws!'

Then the swami of Valaithothu passed on, just as had his counterpart, the man-eater, a few minutes earlier.

The explanation of events seemed simple enough. Due to my cautious movements, the fakir had not known my whereabouts that night. When the man-eater had started calling, somewhere in the direction of the road beyond, the hermit had stolen out of his hut, guided by the tiger's calls, knowing he was safe enough so long as the man-eater kept on roaring and he knew the animal's whereabouts. Undoubtedly he hoped to discover where I was hiding and to practise some form of deception, just as he had succeeded in doing two nights before.

Meanwhile, unknown to him, I had left my machan on the hillside, stolen down to this hut and imitated the tiger's roars three times. The fakir must have been somewhere near the road at that moment. Hearing two separate roarings and believing there were two different tigers, one somewhere before him and one behind, and not knowing which of the two was the man-eater, the undeniably plucky man had at last lost his nerve and hastened back to the safety of his hut, thereby bringing the real man-eater in pursuit and disaster upon himself.

I walked to the settlement, where I found Boora still awake.

'I have shot the tiger,' I announced.

With bated breath, he asked: 'Where is the holy one, *dorai*?'

I shrugged, and sat on his threshold. 'He is dead, too,' I muttered.

Boora sighed, almost in satisfaction. 'Do you now believe?' he asked me.

I laughed, saying, 'The tiger killed him before I could shoot it!'

There was silence for a few seconds as Boora looked at me with widened, incredulous eyes.

'Wh-what?' he mumbled, 'do you mean to say there are two of them; two dead bodies, the swami and the tiger?'

'Certainly,' I replied.

'Are you sure there's not just one dead body—either the tiger, or the holy one?'

'Absolutely certain; just as there are two of us—you and me,' I answered in ruffled tones.

Boora began to call aloud the news to the people in the surrounding huts, and in a minute or two a dozen frightened Irilas stood around us.

'What do you know, brothers,' he announced with pride. 'My *dorai* has shot the tiger and the swamiji!'

I winced, then muttered abruptly, 'Damn it, I told you the tiger had killed the swami and I shot the tiger.'

'Let's see,' they all asked enthusiastically in a chorus.

A little later we were back at the scene of the tragedy, and the excited Irilas almost danced with joy. For their two worst enemies had died that night: the man-eater that ate them, and the fakir who had frightened them almost to death.

When Boora looked down into the still wide-open eyes of the swami, he muttered: 'You—!' and that was all he could say.

Then followed two very bad words.

K. Ullas Karanth

WILLIAM BAZÉ

THE TIGER AT HOME

Indo-China, with its thick and luxuriant vegetation and its savannahs stretching as far as the eye can see, provides ideal country for the tiger and a varied and never-failing food-supply for the creatures on which he lives. The number and diversity of its fauna is a measure of the richness of its virgin forests. Wild cattle and deer roam its immense spaces feeding on the tender leaves or on the seeds of its grasses, according to the season of the year.

In such a land, where wild animals of all kinds are in their element since a kindly nature renews their food supplies at regular intervals, the tiger also comes into his own. His home is in the forest, but at his very doorstep the rich pasturage supplies him with whatever tasty dish he desires.

It is a world in which the seasons regulate and control the lives of the creatures that live in it. When the grass is green the flocks and herds are to be seen there in force. When drought or fire dries up the green pastures they leave them for a time. It is indeed a seasonal migration—and the tiger keeps pace with it.

When the plains have been reduced to ashes by fire their normal inhabitants move off to higher ground until the grass has grown again. Some go farther afield than others. Deer, goats and wild boar are most easily satisfied; they just move by easy stages to the nearest forest land, where they soon find plenty to eat—berries, the bark of certain trees, a few lianas (which they're particularly fond of) and an occasional root growing near the surface. The wild cattle and the gaur[1] are big eaters and more exacting in their requirements. They have to travel long distances to get the amount of food they are used to. The gaur in particular is a tireless traveller. Nothing will stop him and he crosses mountains and valleys, rough and broken plains, deep valleys and raging torrents.

Elephants need their food by the ton, so they have to know in advance where they are making for and follow their traditional migration routes. But their journey is none the less a seasonal affair. They go where they know they will find green food, plodding along at a steady single-minded trot, gleaning here and there a tender shoot or a tasty root, or some giant liana which the other animals have been unable to reach.

The wild buffalo on the other hand is faced with a different problem. He has to find fresh swamps. When the plains are bare of grass he is extremely vulnerable and he has to leave his familiar muddy feeding grounds immediately the reeds have been burnt. He eventually finds some swamp in the forest protected by brambles and spiky thorn bushes. However, like the other animals he has a definite programme which he follows down to the smallest detail.

All the wild beasts, in fact, move off as though obeying a single word of command, and there is nothing haphazard about their migration. It is just one step in the whole system of events which follow in strict chronological order and control the forces of nature as a whole, just one point in the graph of a routine which is as old as time. Century after century the wild beasts have beaten down the same paths, jostling each other in a recurring procession which has never failed.

The tiger—who must follow his 'larder' wherever it goes—is less likely than any other to lose his sense of direction. His compass is his stomach, and it never leads him astray.

He is, by nature, neither a wanderer nor a squatter. If he's anything at all, he's a lazy-bones! But although few people can find much to say in favour of his habits, it must be admitted that the life he is forced to lead is not very enviable either. In all sorts of weather, hot or cold, wet or windy, he must somehow find enough to eat—and not only find it, but catch it and kill it as well. Unkind men are apt to lay in his path tempting morsels which are snares and delusions and the tiger—endowed, fortunately for him, with a highly developed sense of self-preservation—must keep an eternal vigilance against these traps, although handicapped all the time by an exceptionally voracious appetite. And then, just to make life more difficult, his 'larder' suddenly disappears. One day the limitless savannah stretches green to the horizon; the next it is an empty waste of charred and smouldering desolation.

The tiger must adapt himself to his surroundings, especially to the density of the vegetation, wherever he finds himself. He must modify his hunting technique in order to make the best of the conditions, whether for camouflage or for attack and defence. The jungle is full of giant trees, with massive trunks often rising to a most impressive height. There are parts of it where the sun has never penetrated and in these deep fastnesses one moves forward only by cutting one's way. But, despite popular opinion to the contrary, there are often highly nourishing plants growing in some of these thickets, plants which wild deer are especially fond of. These creatures like to hide somewhere near a clearing or in some concealed hiding place, and to avoid the tiger they stay hidden until the evening. But the tiger, always keen to see what's going on, also has a habit of hanging around these clearings.

Now in the open, deers and goats have a good chance of getting away safely from a tiger. They have only to hear, or even to sense, that there is one anywhere near and off they go at a gallop, their hooves beating on the soil with a sound of drums. In the forest, however, it is a very different story.

The bigger animals, like gaurs and elephants, on the other hand, find the forest a pleasant place. They are protected by their very size and by their numbers. Not that this prevents the tiger from taking his occasional tit-bit. If he catches sight of a herd in the distance he will follow it and, like the old highwayman on the coach roads, levy his toll—usually of some careless youngster who has lagged behind.

The clearings in the jungle are places where trees of different sizes are spaced out, or grow in clumps and spinneys, more scattered than in the forest itself. Here the prickly bamboo grows, together with various creepers. The clumps of trees are usually surrounded by patches of grass, green or straw-coloured according to the season. Often, too, there are enormous ant hills, looking from a distance like so many prehistoric monuments. Some of them may be hollowed out at the base where wild animals have licked away the salty earth or where other have sharpened their horns.

In the high plateau of the Moi hinterland, lying in the centre of Indo-China where Vietnam, Cambodia and Laos all meet, the forest clearings are always covered with pines and other conifers. In South Vietnam, in the neighbourhood of Phanrang,[2] the general desolation of the scenery is enhanced by the fantastic bare silhouettes of stunted trees standing up gaunt in the grassy plain. This kind of country crops up again at Son Phan, an area near the border of South Annam which is very rich in game of all kinds, especially the big grey ox and the little red one. In Cambodia these clearings are often visited by the cow-prey[3] and in Laos by a smaller species of gaur. Deer, goats, muntjacs,[4] wild boar and agoutis[5] are usually to be found in the forest clearings; so also is the tiger, who comes here to escape from the wood-leech, elsewhere so abundant. He finds here that his protective colouring, so valuable in the forest itself, is no longer effective and he is forced to move with even more circumspection than usual. He has to watch his prey for hours on end and never lose sight of it. Either he waits in hiding behind some ant hill or some thick bush, or, if the worst comes to the worst, he follows his prey in a favourable wind and leaps upon it in some sunken path when he is least expected.

Whole forests of dwarf bamboo grow in these parts, also a shrub peculiar to the 'red earth' lands of east Cochin China. Stretches of *tranh*[6] alternate with thick clumps and woods of enormous trees growing from the depths of ravines. The dwarf-bamboo forests are quite impenetrable, except on the edge of plantations, where the natives have scratched a field here and there for their crops. Wild deer and boar are plentiful in such places and sometimes steal so much of the harvest before it is gathered that the natives have to call in some hunter to rid them of their troublesome visitors.

The sparse forests of the 'red earth' lands, where the fields of *tranh* seem to stretch to infinity, are the deer's paradise; they are therefore the favourite hunting-ground of the tiger. But for the most part the wild animals seem to have their agreed territories and any infringement on another's recognized preserves leads to somewhat heated explanations.

Another feature of the landscape in Indo-China is the great savannah, where the reeds grow as high as an elephant and are the principal crop. Every year, in January and February, the savannah is swept by great fires, from which it takes a month to recover. Of all the animals that run away to escape the flames, the roebuck is the first to return, followed by the larger deer, then the wild buffalo and the gaur. Elephants, however, wait until the rainy season is well established. Then they arrive in great herds and, in the Lagna[7] plains, they stay from July to December. In the marshes of Bi and of Plao Xieng—two great savannahs in the Moi hinterland—elephants tend to be more suspicious of their trackers—the primitive Phnom tribe—than anywhere else; at Na Kai, however, one of the best hunting areas in Laos, the wide stretches of tall grass seem to have been appropriated by the gaurs.

The tiger adapts his own hunting technique to fit the conditions; as the reeds begin to grow again after the fire, he will avoid any part of the savannah where the cover is inadequate and move to wherever it is thick enough to offer some protection. But as soon as the reeds are fully grown he haunts the paths beaten down by elephants. He creeps stealthily up to the marshland where the small deer may be feeding, then settles down in his hiding place to keep his untiring watch—not forgetting to shift his ground if the wind changes.

But the tiger is as much at home in the forest as he is on the savannah; his only real problem is how to keep in touch with his prey. The feeding habits of the herbivores vary considerably from place to place, according to the apparently casual way in which nature provides an abundance of greenstuff at one spot and nothing at all at another; and the tiger has no option but to follow his victims, wherever they may lead him.

It may seem strange to us here in Europe that every year the great savannahs should be swept bare by these fires, but it is difficult to see how else the greater part of the animal population could continue to survive. These fires, occurring at just the right moment, destroy the dead and yellowing vegetation when it is of no more use, and from its life-giving ashes the new life arises, green and appetizing. The fires start spontaneously at the height of the dry season, but they are stopped by the general moisture of the marshland as they approach the forest areas, so that there is little danger of the great conflagrations which occur from time to time in other parts of the world.

The occasional tiger which tackles a herd of wild oxen on the march and stalks it on its way to new feeding grounds, is usually one that has itself been driven from the savannah by the fires. He might equally have picked on a herd of deer or of wild boar. If any of these creatures should cross his path the tiger will not miss the opportunity for a change of diet. Even if the herd breaks into a gallop the tiger usually succeeds in getting one or two of them. Meanwhile the rest of the herd will be putting on a spurt in their haste to reach a less vulnerable part of the forest, although they may be waylaid again before they reach it. However, nature preserves the balance and the casualties will be replaced by young ones born during the course of the year, so that the size of the herd is about the same on every migration.

A good supply of drinking water is one of the essentials for the migrating herds and will partly determine the site of both temporary and permanent settlements. If a tiger discovers an area which is rich in the kind of food he enjoys he will plan his life to get the best out of it for as long as it lasts and move off only when its stores are running low.

When he is on the move the tiger takes every possible precaution to avoid being caught unawares. For his home he likes a thick dark wood, guarded by an impenetrable screen of undergrowth, and anywhere with a narrow winding entrance is no use to him. Above all he must have peace and quiet to digest his food, so he will always provide himself from the first with every strategic advantage and ensure a quick get-away at a moment's notice if necessary.

He drags the bodies of his victims as far as possible into the undergrowth, beyond the greedy prying eyes of the vultures; but he always keeps open an escape route in case he should be surprised while he is feeding. Naturally he is happiest if he can make his kill when he is quite near his lair.

Things are more difficult for him in the clearings. There he must cover his 'larder' with anything handy and retire—perhaps quite a considerable distance—to wherever the conditions provide a really safe hide-out. This means a good walk when he comes out at night to resume his feast, which can be inconvenient, to say the least. There is always the chance that he may return to find some uninvited guest helping himself to his dinner. Any 'temporary' lair is abandoned as soon as the corpse is all eaten, which may take several days, as the tiger is very strongly attached to his permanent quarters and uses them as his base for his foraging expeditions.

One part of the country I knew very well was Co Hen, a remote district, very rich in tigers, lying between the last of the Lagna savannahs and the first outcrops of the Annamese mountains, where the tigers were well settled in. It was a delightful spot. The forests in the dry season and the plains in the rainy season were full of roots, grasses, fruits and young shoots of all kinds. In a comparatively small area you would find wild cattle, stags, small deer and wild pig in herds anything from fifteen or twenty to a hundred or two hundred strong. I used to camp out near a Moi village and after one night's tour of inspection of the village allotments with my hunting lamp, I was able to do the people a double service; I would save the rest of their crops and at the same time restock their larders with fresh meat.

Near Co Hen there was a special piece of forest I shall always remember. It was not very extensive but there was a clear stream running through it. Right in the middle was a magnificent natural retreat for tigers, and I never knew it to be unoccupied. As soon as I polished off the sitting tenant, so to speak—or at any rate the first tenant I had encountered on the site—an even finer specimen would appear. Having learnt somehow or other that the property was vacant he had come to take possession without worrying over much about any legal preliminaries. I would remind the newcomer—by the same reception I had accorded his predecessor—that there were certain decencies to be observed, when along would come a third squatter, anxious to take up residence without further delay. This remote and unsuspected corner, in fact, must have seemed to the tigers of the district a perfect paradise—plenty to eat, running water to drink and, until I arrived, absolute peace and quiet.

But I also had an interest in this desirable residence. . . . In the course of one year I killed six tigers, including a pair I surprised in the middle of a romance. All six of them came in turn to the bait I had set and I shot them all from the same 'hide'. The only variation was in the time: the lovers were late—or perhaps something had happened to upset my usual time-table that night.

The death of the loving couple put an end to the series, and nearly put an end to me. I had shot the male and was moving cautiously towards him. He was not quite dead and, with my finger on the trigger of my gun, I stood there waiting. The end came with a great hoarse death-rattle—was it a cry of agony or a last appeal for help? At the same moment I heard growlings behind me. Some instinctive reflex made me turn round sharply, and there, coming straight for me with her belly close to the ground, was an angry tigress—the widow, no doubt. I gave her a charge of buckshot from a range of five yards, which made her rise on her hind legs to her full height like a bear. So I gave her my second charge of shot, full in the chest this time. She fell backwards and rolled over and over, while the whole forest seemed to re-echo with her roaring.

I was a bit shaken, but I reloaded my rifle with two more cartridges. However, two rounds of buckshot at four or five yards range had taken care of the tigress. What puzzled me most at the time, in fact, was that death had not been instantaneous at so short a range. I discovered later that my cartridges were badly packed.

Since the war it has not been possible for me to go back to Co Hen, much as I should have liked to. Yet I can still see, quite clearly and without any effort, the spot where I used to camp, the paths I used to and from my 'hides' and the clear flames of my brazier trembling in the cold damp air of the forest night. And still the sounds come back to me from those happy years—so recent and yet so long, long ago—of animals calling to each other through the darkness in their fear, or fleeing terror-stricken at the approach of danger. I hear the strange metallic cry of the tiger, answered by the thunderous blare of the elephants disturbed at their evening meal.

Memories come crowding in on me. Above the deep shadows of the forest the tropic sky is filled with stars; millions of fire-flies light up the dark foliage; and as I sweep the grass with my lantern I see the unnumbered shining eyes of the little deer as they slip silently past to their feedings grounds.

There are the stars which are of the heavens . . . and there are the stars which are the earth's.

* * *

THE DEAD-BAIT LURE

Provided you have the carcass of one of the larger animals, and provided it has reached an advanced state of decomposition, there is no doubt that the baited trap is easily the most efficacious method of getting the tiger where you want him. The carcass can be smelt a long way off and never fails to attract a tiger; and there is enough food there to last him sometimes as long as a whole week if he is not disturbed.

From the moment he has touched the bait the hunt is on, but only if it is dead bait. 'Live' bait in my experience often lets you down. You can tie up a cow, a horse or a buffalo—if you are sufficiently hard-hearted—and if it does happen to attract a tiger it will only be by accident; it may stay there indefinitely without being disturbed, and if the wind is in the wrong direction and the bait happens to be behind a bush, a tiger will pass within a few feet of it without suspecting its presence. I have often seen a tiger's footprints in the morning quite close to a tethered animal, although the hunters had been keeping watch all night without having seen one pass.

In 1921, I bought an ox which had dislocated a shoulder in an accident and I thought it might make an interesting bait. It was in the care of an old herdsman who led it out to have a good feed and a long drink and then took it to the spot where it was to act as bait. A lookout was made in a neighbouring tree and I was able to watch proceedings. I sat in that tree for weeks and nothing happened. Over and over again the ox was moved but all I got from the experiment was an attack of malaria. The beast himself, tired of waiting, would lie down every evening when it got dark and chew the cud quietly until sunrise. One night while I was watching, a man hunting by lantern came past and shot the poor ox. Realizing his mistake he made off quickly, and my mobile bait, which had failed to interest a single tiger, brought joy to the hearts of the natives, who cut themselves some first-class steaks from the carcass.

The deceased was replaced by a sick buffalo, but it died after three weeks without even a formal courtesy call from a tiger. As soon as it began to go rotten however, the tiger appeared. Unfortunately this was before I knew about the powerful attraction—for a tiger—of a decaying carcass and I did not trouble to keep up my vigil after the buffalo died. Hitherto all the tigers I had killed had either been attacking my livestock or had crossed my path at night while I was hunting deer.

But my education on this point was soon carried a stage further. In the same year, 1921, a famous hunter, Fernand Millet, developed the 'dead bait' system—of which, in fact, he was certainly the inventor. I was one of the first to use the new technique, and my bag of tigers of all sizes was pleasantly enriched as a result. It was really a revolution in big-game hunting. Fernand Millet himself, before he introduced the new method, had met with disappointment after disappointment and had spent many *nuits blanches* in his 'hides' watching a tethered bait. One day he discovered that when a tiger killed his prey he would devour every scrap of the corpse whatever its state of putrefaction. He told me how he had shot an elephant in the forest and learnt later that the great carcass had provided a tiger with meals for a week. He went to see what was left of it one evening at dusk, found the tiger there and shot it.

Once he discovered what an attraction putrid meat has for the tiger, he came to realize that so long as the tiger's meals are guaranteed he gives no thought for the morrow. Since laziness is his besetting sin, he has no objection to an existence which required no effort and he does not think of looking for a fresh victim until his reserves are completely exhausted.

The first essential for a hunter who proposes to use this technique is a detailed knowledge of his district. For one thing he must set up his watching post in such a position that the pestilential odour of the carcass will not make him sick. Secondly he must be careful to keep the approaches to the bait clear and leave adequate access for himself to and from his lookout post. The bigger the carcass the wider it will broadcast—given a favourable wind—its delectable invitation to the tiger. The 'Aristotle's' deer and the wild boar make good bait, especially if there are two or three carcasses available. Best of all are the greater oxen, gaurs and bantengs, wild buffalo and, of course, the elephant. Unfortunately these giants are not easily transported any considerable distance from the spot where they are killed to the best place for putting them down as bait. But the difficulties are not insurmountable. With a local cart and a few draught buffalo it is possible to shift large pieces of gaur or elephant and pile them up together to increase the size of the bait, although this manoeuvre is naturally more difficult in hilly country.

The best plan is to buy a buffalo from the nearest village—or, failing a buffalo, an ox or a cow—and slaughter it on the spot where it is to be used.

Ordinary deer make poor bait, partly because they are not large enough and partly because they are comparatively easy to dismember. The tiger snatches a limb or a joint and runs off with it to eat it under cover somewhere. He does this again and again, with such speed that the hunter, although he is on the look-out, seldom has time to get in a successful shot. The flesh of the larger animals is tougher and the tiger has to eat it where it lies, at the same time offering a steadier target. The flesh of the domestic buffalo is particularly tough; that of the domestic ox and of the horse less so, but still by no means tender.

Choosing a site for the bait is a matter for careful consideration. It must be somewhere in the jungle, preferably near a stream, so that the tiger can wash his food down every now and then. If you put a carcass anywhere in the open it will be cleared away in no time by vultures. Moreover in open country the tiger usually waits until nightfall before coming to dinner. If he is hiding among the trees on the plain he may approach the bait from any direction, although the watchful hunter can usually see him coming.

The ideal place for a bait is somewhere on the skirts of the forest where the tiger is known to be hiding. It is out of sight of the vultures, and the hunter can approach his lookout post from the opposite direction to the tiger's.

When the experienced hunter has selected the ideal spot he can then prepare an approach route for himself by cutting down the long grass so that he can creep up silently and surprise the tiger while eating. If he builds a protective screen about twelve or fifteen yards from the bait, the hunter will get a splendid view of the proceedings. Sometimes a tiger—startled despite the hunter's careful precautions—will put on a noisy act and leap around in a narrow radius instead of running away, as one would expect him to. When this happens the sport becomes more interesting—but it can be a bit of an ordeal for a beginner.

A point worth watching is that the bait must not be laid in such a position that the tiger can take cover behind it. It is useless, for example, to fasten it to a big tree; the tiger will simply help himself to a joint and retire behind the tree to eat it. If it is tied very loosely to a stump it is not so bad because the tiger is bound to remain visible. The rattan cane, which grows everywhere in the forest, is ideal for tying the bait as it is very pliant and will withstand any amount of pulling and twisting where wire might tend to snap. Any kind of chain makes too much noise and arouses the tiger's suspicions.

Of the various types of 'hide' the best is a look-out in a tree, preferably behind a screen of vegetation. A professional hunter of pre-war days, M. Plas, invented a look-out built actually on the ground and had his best results with it. It was a kind of artificial bush, just big enough to allow a man to sit inside it on a very low seat. It offered little protection against rain, but it was a good idea because it provided perfect camouflage. From its depths the hunter could watch the bait through a spy-hole cut at eye-level, and in due course he would use the same hole for the barrel of his gun.

This type of lookout enabled M. Plas to bag the most suspicious tigers—those which had been shot at and wounded on some previous occasion and would abandon the bait the moment they spotted a 'hide', whether it was behind a screen or simply built in the main branches of a neighbouring tree. Over and over again I have seen the suspicious type of tiger assuring himself that the bait is not overlooked by any possible lookout post. What is more, they have often played hide-and-seek with me for quite a while before giving me a chance to shoot. They would fix it so that they could see me without being seen and just to prove they had seen me they would always come and have a good go at the bait during my absence. But I usually knew a trick as good as theirs. Sometimes I would build a fresh lookout on the opposite side of the bait; sometimes I would build one on the ground in a big bush; sometimes I would watch for them a long way from the bait and surprise them on their favourite route to it; and sometimes I would wait at the spot I knew they would visit for a drink after their meal. Whichever way I chose, I usually got them in the end.

The most cunning of them cost me several baits before they finally stopped a shot from my rifle. Four times tigers broke through a cleverly devised ambush but were subsequently caught in a steel trap. Each time, on returning from the bait, I found the tiger firmly held by a hind foot and reduced to impotence—but not to silence. There was nothing he could do about it but roar with anger.

* * *

To sum up, then, one needs a fairly shady spot, with an inconspicuous approach, and the ground all round the bait should be cleared of everything which may obstruct the hunter's view. If the tiger is clearly visible from the moment he appears on the scene until the moment he begins to feed, the hunter's problems are enormously simplified. It is indeed possible to bag a tiger without taking all these precautions, but many an inexperienced hunter has come home empty-handed simply because he has not done everything he might have done to guarantee success. Others have chosen their spot well and laid out their bait but have omitted that preliminary preparation of the site which is always such an advantage. If the clearing of the site and the construction of the look-out are left too late the tiger's suspicions are bound to be aroused and he will go and seek his dinner elsewhere.

Every hunter has his own ideas about a look-out. For myself I make no bones about constructing a really solid platform, with the logs well spaced to cut out all risk of creaking. It has three sections; the one at the back is protected from the rain and the one at the front is open to the sky. In the middle is a comfortable seat with a back, which allows me to sit in a normal position and not get cramp after sitting still for long periods. On my chair is a camp cushion and under it I keep a thermos, provisions, first-aid equipment and some dry clothes. The whole contraption is concealed by leaves and branches, with a number of gaps through which I can easily watch the arrival of the tiger and his subsequent movements. The platform is covered with a layer of sacks to make a thick carpet which effectively deadens the sound of my footsteps. I even have a primitive rack for my guns and a projector mounted on a branch, which my assistant operates on my instructions as soon as it is dark. So I have the choice of awaiting the tiger on the ground behind the screen, or of sitting up in my look-out to keep my vigil. I find the latter method specially welcome in wet weather.

Notes:

1. A giant ox with smooth hair and magnificent horns, which somewhat resembles the aurochs. It is dark chestnut—almost black—in colour,

with white 'stockings' reaching almost to the knees. Some specimens grow to a height of 6 feet or more at the withers.

2. A town near the sea and about 90 miles from the frontiers of the old Indo-China.
3. A black ox found in Cambodia and Laos which is peculiar in its habit of keeping the points of its horns 'splintered'—some, in fact, seem to be wearing a bunch of feathers on their horns. This peculiarity is due to the fact that the creatures sharpen their horns by digging them into the ground, and the ground thereabouts is exceptionally hard.
4. A small reddish goat which is peculiar in that it has movable horns with which it has been known to cut a dog's head clean off. The cry of the muntjac is a sort of raucous bark.
5. A reddish-brown mammal found in America and Indo-China. The name is given also to a small deer, about the size of a wild rabbit.
6. A tall grass, sometimes called *paillotte*. The natives use it as straw for thatching their cottages.
7. The great plain of eastern Cochin China, over 250 square miles in extent, crossed by the Lagna river and its many tributaries. It is the most beautiful of all the National Game Reserves, and when you fly over it you see great herds of elephants and wild buffalo, frightened by the noise of the engines, charging in all directions.

DIET, KILLS AND AREA COVERED

The Malayan tiger does not have such a varied diet as the tiger of India where ungulates—serow, goral, nilghai, black buck, swamp deer, chinkara, sambur, chital, barking deer, four-horned antelope and mousedeer—are plentiful, in addition to various species of pigs and monkeys, hares, rabbits and an assortment of the bigger birds. In Malaya, the tiger must content himself with pigs, which appear more frequently on his menu than anything else, with an occasional *rusa* (the Indian sambur) and a still more occasional barking deer. Mousedeer, which are not much larger than hares, are not big enough to make a satisfying meal. Nor are the peacock or jungle fowl, although any one of them would be most acceptable to a really hungry animal. I have never heard of a Malayan tiger attacking a wild elephant, although such encounters have been reported from India. In his book *Big Game of Malaya*, E.C. Foenander tells of a *seladang* (the Malayan wild oxen *Bibos Gaurus Hubbacki*) calf being killed by a tiger, which was then prevented from feeding from it by an encircling movement maintained by adult members of the herd.

Doubt is expressed in the same book about the ability of a tiger to break the neck of a mature bull *seladang*. While I agree that a tiger would need to be very hard-pressed for food to tackle so formidable an adversary, I must point out that I have seen several of the huge bull Indian water buffaloes, used by the Malays to draw wood or sled-like conveyances carrying fruit, etc., which have been killed by tigers. These big bulls cannot be so very far behind a *seladang* in size, weight and strength, although they may lack the wild animal's ferocity. The important thing to remember is that a tiger does not break his victim's neck by force, but kills by biting into the neck near its junction with the head, a function for which the long canine teeth are admirably suited.

Although I frequently came across the tracks of tapirs (*Tapiris malayanus*) in Trengganu, I never heard of one of these odd looking and almost defenceless animals being killed by a tiger. However, in the June 1950 issue of *The Malayan Nature Journal*, J.A. Hislop included in an article on these animals a note by C.E. Jackson describing how a tiger had attacked a tapir at some waterworks near Kuala Lumpur, both beasts having fallen into the 'intake well'. The tiger escaped, but when the tapir was hauled out after considerable difficulty it was found to have suffered severely from its attacker's claws. Its lower left jaw-bone had also been completely crushed by a bite. The animal was accordingly destroyed.

It may be of interest here to record that tapirs are not as mute as many believe. I have heard them in the late evening making a noise remarkably like the persistent shrill yapping of an excited dog. The sound would have passed unidentified had I not realized that there could be no dogs in that particular area and questioned the local Malays who were with me. If the tapir is attacked, it would also be reasonable to suppose that the Sumatran rhinoceros (*Rhinoceros sumatrensis*), which exists in Malaya in rapidly diminishing numbers, must also occasionally fall victim to a tiger when immature. These animals are met with so infrequently (I knew definitely of only one pair in Trengganu) that the average tiger would probably pause in amazement at the sight of one instead of attacking it.

Cases also occur on the sea coast in which tigers kill the giant turtles as they come up the beach to lay their eggs. The turtle is turned over onto its back and thus rendered helpless. It is then killed and eaten at leisure.

Collecting these turtles' eggs, over a hundred of which may be produced by one reptile in a night, is a profitable occupation in parts of Malaya. Large sums are paid to the State Government concerned for the right to collect all that are found along each section of the shore. The Malays engaged in gathering this harvest from the sea are in the habit of spending the night on the beach to mark the spots where the females deposit the eggs and to ensure that they are not stolen. The fires which they light for warmth and companionship may be seen burning along the coast until late at night. They are apt to die out before dawn, however, for by then most of the watchers are sound asleep, wrapped from head to foot in *sarongs* to keep off the sand-flies.

It was to such an East Coast beach that a young tiger came near dawn one morning in 1951. There were the dying embers of a fire. Beside them was lying a shape the like of which he had never seen before. It had no head. It appeared to have no limbs. It did not move. Curiosity overcame him. He moved closer. There was still no sign of life. Stepping up to it, the tiger gave the shape a gentle, playful nip to see what would happen. The roar of pain and terror which resulted so startled the tiger that he made off hurriedly. A few hours later, the Malay was in the Kuala Trengganu hospital, the bones of one elbow crushed and splintered by that one exploratory bite.

Apart from wild animals, the Malayan tiger which is prepared to enter or approach a village has buffaloes, cows, occasional goats, sheep, pigs kept by Chinese, dogs, cats, tame monkeys and poultry on which to feed. Malayan cattle are very small compared with European farm types, the exception being the big Brahmini bulls usually owned by Indians. Dogs are taken frequently; cattle run them a close second, losses among them being high because of their curiosity. I have never heard of a tiger killing two buffaloes one after the other, but the death of two, sometimes three, cows in one encounter is not unusual. What happens is that the herd of cows is grazing in daylight, or lying down, when the tiger attacks the animal which he has singled out. The remainder flee for dear life, tails held high as they gallop off. Then they all stop to look back. Seeing nothing, one or two return to find out what has happened. They approach too close to the tiger. Enraged at being disturbed over a fresh kill, the tiger attacks again and another cow dies, quite unnecessarily. Tigers will eat almost anything that they can kill or which they find lying dead. It is known that crabs have been eaten as well as frogs.

* * *

FEEDING

Once the kill is in a spot to his liking, the tiger begins to feed. In nineteen cases out of twenty, the buttocks are eaten first, then the hind legs and so on up the body towards the head. A half-eaten kill left by a tiger is nearly always the same in appearance, consisting as it does of the forward part of the body, the front legs and the head. If the kill is a cow or female buffalo in calf the foetus may be torn out and eaten first. Leopards and black panthers on the other hand nearly always feed first from the stomach. Some tigers will remove parts of their kill and bury them before they begin to feed, this being most common with the trotters of wild pigs. These are severed from the legs and concealed in the ground, some pains being taken to pat down the covering earth. Other tigers acquire the habit of hiding a half-eaten kill beneath leaves and grass, this being done so neatly that it is difficult to believe that an animal has collected and arranged the covering materials. The entrails of a kill are never eaten by a tiger. They, and many other small pieces of flesh, bone and skin, are dragged off by the monitor lizards which, attracted by the smell, visit the kill when the tiger is not there.

A fully-grown tiger will eat up to 40 lbs. of meat at a meal, depending on how hungry he is. It makes no difference to him how putrid the flesh has become. I have heard a tigress making unusual blowing noises when eating from a dead buffalo of which little more was left than a seething mass of maggots and have no doubt that she was blowing to clear the grubs from her nostrils. Hungry tigers are far from being particular feeders and will readily devour any carrion to which they have been attracted by smell. Sometimes they stoop to outright scavenging, as did the tigress and her cub which made a habit of visiting a town refuse dump in Trengganu to see what scraps they could pick up.

A tame animal which has died a natural death, or a wild one that has been shot and put out as bait, proves just as acceptable to a tiger as a kill that he has obtained himself—a fact which enabled me to end the career of more than one cattle thief. Distinctive and repulsive noises are made while feeding, including the sound of the meat being torn off and swallowed in great gulps. A close examination of a dead buffalo, from which a tiger had been permitted to take two mouthfuls before being shot, disclosed that nearly 4 lbs. of meat had been consumed in those two bites. Grinding of the teeth, accompanied by a soft rasping noise from the throat, is frequently heard as a tiger returns to a kill. This is, perhaps, the feline version of lip-smacking in anticipation of a meal. The noise produced from the throat is much like that made by a domestic cat when it sees a bird beyond its reach, except that it is considerably louder.

It is unusual for a tiger not to return to his kill, once he has fed from it, but cases must sometimes occur when fresh food is almost forced upon him. A blundering pig might approach too near a tiger's lying-up place, for example, and awaken the instinct to kill. The Kemasek man-eater, before he began the final series of killings which made him so feared in 1951, behaved in a most peculiar manner on discovering a live cow which I had tied up for him. He was walking along a sandy crack carrying a dead mousedeer in his mouth when he heard or smelt the cow. He could not have been hungry or he would have eaten the mousedeer instead of carrying it. Evidently he had already eaten that night and had come upon the mousedeer unexpectedly. The cow must have appeared as another easy victim, not to be passed by although he had no need for it. Tossing the mousedeer to one side (I found it five yards from his pug marks), he killed the cow and dragged it off into the nearby swamp as I had intended he should. Having killed the cow he did not know what to do with it and contented himself by chewing most of the flesh from the face instead of eating the more succulent parts. He then left both mousedeer and cow and walked off, not bothering to return on either of the two following nights on which I waited up for him.

Although a tiger almost invariably returns to a kill from which he has fed, he will hardly ever come back to a dead animal which he has left uneaten of his own accord. It is hard to account for these apparently aimless kills when the tiger goes to a great deal of trouble to secure them, but the fact remains that such instances do occur. There is no apparent reason why the kill is left. I have taken great pains to ensure that cows killed in this way have not been disturbed, and have checked by the absence of footprints afterwards that no one went near them, but the tiger seemed to have lost all interest in them.

I do not include in this category kills over which an attempt has been made to shoot the tiger, as there is always the possibility that, however carefully the preparations have been made, something has been done to alarm it. Cattle will be left lying in the open after being killed in the daytime, but that is because circumstances combine to make it inconvenient for the tiger to drag them away. Next morning it is usually found that the tiger has returned and made off with them. It is only when the tiger has a completely free choice in the matter and elects to leave his kill uneaten that he can be depended upon never to return to it again. I have known other tigers to be attracted to such kills, however.

Tigers are certainly individualistic. One that I knew of always made for the nearest water with his kill and, having reached it, proceeded to tow the dead animal along the shallow parts. This must have simplified his task of removing his kill, but added considerably to mine when I had to try and follow him up, as there was a complete absence of pugs or drag marks.

Contrary to popular belief, tigers do not have lairs, although they will occasionally carry off their prey to the same spot when no alternative feeding place is available. A case in point was a tigress which made off with a number of dogs from dwellings close to a police enclosure. This animal took the dead dogs up a steep hill behind the police station and ate them among the bushes on its summit. She had no option unless she was prepared to walk through the streets of the township with a dead dog in her mouth. When I found the spot, I collected from it six dog collars, two of which were later identified by the animals' owners. The feeding place was so close to the police station that I could look down from it on to the roofs below. I wondered, as I stood there, how the Malay policemen and their families would have reacted had they known how near a tiger had often been to them.

AREA COVERED

Tigers cannot normally afford to remain in one locality for more than a day or two at a time, because local game will begin to move out of the area almost as soon as a tiger moves into it. Tigers have a pronounced musk-like smell, easily identified by other animals from a distance. The noises which accompany a kill, such as the shrill squeals of a stricken pig, provide further warnings to game to make itself scarce. I knew of one isolated rubber plantation which could usually be depended upon to yield a wild pig for the table, if one knew how to stalk these watchful animals. Time and time again Pa Mat and I found, on the narrow path leading to this spot, the pugs of an old tigress. If they had been made within the last two days, we knew that our chances of shooting a pig were slender. In some of our other haunts, too, we noticed that if there were fresh pugs there would be few new pig tracks and vice versa.

It is for this reason that tigers do not stay in one place for long, but must be on the move continuously. Their movements, plotted over a period of months, produce definite patterns. The most common of these is a rough triangle, the three points of which are fluid but are generally five to eight miles apart. Having secured a victim at one point of the triangle, a tiger might remain there for one, two or three nights, until he has eaten out his kill. He then moves on to the second point and will remain there for another two or three days, before transferring to the third point. Another short stay there and he returns to the vicinity of his starting point. Such a programme not only enables each hunting area to be visited every nine to twelve days, but also permits the tiger to follow up the game which tends to leave each part of his area as he visits it.

Pa Mat and I proved many times that a tiger which had secured a kill, whether from village livestock or from wild game, could generally be expected to return to the neighbourhood within about ten days. Often the circle, or triangle, is completed with monotonous regularity. Occasionally the tour is accelerated or slowed down, depending on the success of the tiger's hunting. Sometimes tigers will go off on long jaunts. I once followed the pugs of a male from the point where I found them behind my house. The tracks led across country to the main East Coast Road and then kept to the roadside until approaching the river ferry, a mile from my home. Here they turned off so that the tiger could swim the river a safe distance up stream from the building occupied by the Malays who operated the ferry. I lost the tracks there, but the pugs appeared on the roadside again five hundred yards beyond the river and continued to keep to the road, apart from a few deviations, until they branched off into the jungle near Panchor. This tiger covered not less than twelve miles that night from the time that he passed behind my house. How far he had gone before that is not known. The same pugs were seen on this route once or twice again. Then I found them no more. Later I discovered that this animal had apparently been on a visit of inspection to an area left vacant by the death of another tiger which I had had to shoot. Finding it more to his liking than his own, he had eventually moved into it. It is one of the surprising facts about tigers that when one is killed another soon appears to take over the unoccupied hunting 'block'.

Once he has adopted a particular area as his own, a tiger will strenuously resist any attempt on the part of another tiger to poach upon his preserve. This does not often happen, but when it does there is usually a fight to decide which animal shall remain in the disputed territory. These are, I believe, the only occasions on which fights occur except during the breeding season. A male does not appear to resent the presence of a tigress in his area, especially if she has young with her, but they do not consort or share kills, preferring to keep out of each other's way. I learnt of one battle royal between two males in Trengganu in 1951. Local Malays heard the sounds of the fight coming from the top of a hill during the middle of the morning. The struggle went on during the day, the scene of it slowly moving down the hill on to the banks of a river. Here, at about four o'clock in the afternoon, the younger and smaller animal succeeded in overcoming and killing his opponent. Tiger-like he could not resist a brief feed from the body of his fallen foe before leaving it.

These fights are rather like a quarrel between two domestic cats, in that far more of the time occupied is spent in circling, snarling and threatening than in coming to grips. The exchanges themselves are of short duration and are divided by long periods of manoeuvring. Although I have not seen one of these fights myself, I have shot a tiger which had recently taken part in one. The most obvious of this animal's injuries were a deep bite on one foreleg, the absence of two claws from the opposite forepaw and damage to the mouth, which included two broken canines. When skinned this tiger was found to be so full of black thorns that barely an inch of the underside of the skin did not have its share of them.

Unlike domestic cats and most other felines, including leopards, tigers have not the slightest aversion to water. On the contrary, they like to lie in it during the heat of the day and are very strong swimmers. One is reputed, before the war, to have been in the habit of swimming some four miles to an island off the coast of Malaya to hunt the pigs which existed in numbers there. Malays will tell you that when a tiger sets out to cross a river he makes for some fixed point on the opposite bank. If the current carries him downstream, or if he misses the mark for some other reason, he will return for another attempt rather than land at any other place than that which he has chosen. This was another Malay belief that I could not readily accept, but one day, when I was in the jungle during floods caused by the monsoon, I saw something which made me wonder whether it was true or not.

I was with a police patrol. We had followed a track which led us to a swollen, fast-moving river which it was necessary to cross. Along this path were the pugs of a tiger, a fine male from the size of them, which led straight to the water's edge and there ended abruptly. Boats were procured to take us over the river, the crossing being accomplished with difficulty because of the swirling, eddying flood of water. I was among the first to cross and there were the pugs of the tiger where he had stepped ashore on the exact centre of the continuation of the track and had gone on his way. The volume of water was so great that even so powerful a swimmer as a tiger must have been swept some distance downstream. Yet he had obviously swum back, probably along the slack water against the far bank, to land where he had originally intended.

One must assume that sometimes during floods of this nature a tiger must be swept so far downstream that he is lucky to reach the further bank at all. He would then have no option but to swallow his pride. They must also be drowned at times too, for the first dead Malayan tiger that I ever saw was one floating down a river after severe flooding. Tigers do not as a rule climb trees but have been known to ascend to a considerable height when the danger which prompted them to climb was sufficiently grave. I know that at least one Malay has lost his life through depending on the fact that he could climb and that the man-eater he was hunting could not. Chased by the animal, he struggled up a sloping tree with a sense of relief which turned to panic when he found the tiger following him. He could not descend because the tiger was below him, so he had no option but to jump for it. The tiger followed suit and caught him as he reached the ground.

One of the differences between Indian and Malayan tigers is that the latter, when walking at their normal speed, do not normally place their hind feet on top of the pugs made by the front feet. Both the Indian tiger and leopard frequently do so. I have seen a great many tigers' pugs in Malaya, and not a few of the pugs of leopards, but the only time when they overlapped was when a tigress was stalking a dog across a newly-dug garden. In this case, I believe, the animal deliberately placed her back feet in the deep depressions caused by her fore-paws so that she could move more steadily and silently.

Thc pugs of the male tiger reproduced on facing p. 49 are typical of those seen in Malaya. This photograph, taken of pugs on dew-marked silver sand, also illustrates dearly the difference in size between the fore and hind feet, the latter being appreciably smaller. This difference in size is sufficiently marked to lead inexperienced Malays and Chinese to believe that two animals, one bigger than the other, have left the tracks. Such reports of two tigers having passed along a track or through a village often reached me.

The Indian tiger frequently injures itself by impatient attempts to kill porcupines, the quills of which penetrate deeply and cannot be plucked out. They are bitten off, but cause suppurating wounds which in some cases have incapacitated the animal sufficiently to cause it to turn man-eater. I, personally, have never heard of a Malayan tiger killing a porcupine although this must sometimes occur. Of the thirty or so dead tigers which I have examined I have never seen one carrying either the marks of porcupine quills or the quills themselves.

KESRI SINGH

LIONS AND TIGERS

The Indian lion is now so rare that its very existence is often overlooked. But, historically speaking, it is not long since lions were widely distributed over the peninsula and probably fairly common in the jungle country of what is now Rajasthan. Today wild lions are found nowhere in the world outside Africa save in the Gir Forest in Saurashtra, where the sole survivors of the Asiatic lions that once ranged not only over India but through Persia and Mesopotamia to Europe, leaving an indelible mark on the art and ideas of innumerable civilizations, have found a last asylum.

Whatever may have been the case further west, I do not myself think that the gradual disappearance of the Indian lion has been mainly due to human intervention. I believe the tiger to be responsible. Lions and tigers in the same area are direct competitors for exactly the same prey, and tigers are semi-solitary by habit, jealous of their hunting ranges, and notoriously intolerant of any sort of rivalry. They are, as a general rule, more powerful, more cunning, and better equipped for life in the jungle.

Tigers seem to have come into India from the east, from China via Burma and Assam to Bengal from whence they spread out over the country, driving away or exterminating the lions wherever they encountered them. It is noteworthy that the lions' remaining stronghold in Gir is in an isolated area separated by more than a hundred miles of difficult terrain from the hills that constitute their rivals' nearest territory.

On three separate occasions when arranging public beast-fights I have put a lion and tiger into the arena together. In each instance the result was the same. The lion attacked and soon got the worst of the encounter. After one or two blows from the tiger's forepaws it would retire, and since the tiger never followed-up the performance would be over. However, it cannot be argued from this that the tiger would not, in a state of nature, take the offensive. Wild tigers that have been captured and put in a ring are usually most reluctant to attack any other large creature—except, very reasonably, a man—in front of spectators. But in the jungle they fight each other ferociously, as their wounds testify.

The late Maharajah of Gwalior, believing that lions once roamed over the territory of his state, determined to try and re-introduce them, and with this object in view imported three pairs of fine African lions. The area selected for the experiment was the Sheopur-Shivepuri forest range which covers an extent of over one thousand square miles.

I have already referred to the great enclosure which was made at Dobe Kund to accommodate these visitors and allow them to acclimatize themselves. It was situated in a peculiarly wild and lonely spot, half-way between Sheopur and Shivepuri, and consisted of a large area divided internally and surrounded by a stone wall twenty feet high broken by strong gates, through which, at intervals, buffaloes were driven. The object of feeding the lions on live meat in this way was to prevent them losing the ability to kill their own food. In fact, they not only became acclimatized, but nourished and bred in the enclosure.

The surrounding forest abounded in tigers who seemed to be attracted by the roaring of the lions on the other side of the high wall. Twice while making one of my periodical inspections of the place I found a tiger prowling near—probably anxious to challenge the unfamiliar voices.

After about four years it was decided to start releasing the lions. They were not all to be let out at once, but a pair at a time. The first couple to go disappeared without trouble, simply vanishing away into the wilderness. But the second couple, freed not long afterwards, quickly returned to the neighbourhood of the enclosure where they caused great alarm to the shikaris who brought supplies for the captive lions within. For they hung around near the track up which the buffaloes were driven and at a suitable moment charged one, killed it, and started feeding there and then on the carcass. Fortunately they did not attack any of the men.

The morning after getting the news I went to Dobe Kund with a party of shikaris and beaters, and with a good deal of noise drove both lions away from the cover where they were lying up. My hope was that they would be sufficiently upset by the experience to clear out altogether. Shortly afterwards however, some of the shikaris, who were responsible for taking up the buffaloes, found a trail that led them to the body of the male lion. He had been severely mauled, and since we knew that the first pair of lions had left the area it was morally certain his death was the work of a tiger. There was no other creature capable of inflicting such wounds. What had no doubt happened was that the pair on being driven from the immediate vicinity of the cattle trail had invaded the hunting ground of a tiger. The lion had stayed to fight and be killed while the lioness had made her escape.

Owing to breeding there were eight more lions in the enclosure. Six of them were released, two at a time, without incident. Then finally, after a further two months' wait, the turn of the last pair came. These were the only ones that gave any real trouble, but they soon made up for all the rest.

The Sheopur-Shivepuri forest range had been chosen partly because it was a wild and very thinly populated area. There were no proper village communities, but here and there a few meagre, scattered hamlets. Soon after the last two lions had been liberated reports that they had taken to cattle lifting began to come in. The inhabitants of the hamlets were poor and possessed no firearms, and the lions would simply emerge from the jungle, seize one of the beasts grazing by the huts, and devour it more or less on the premises. To keep them off the inhabitants strengthened their thorn fences and kept their cattle as much as they could behind them. The policy worked for a time, but after a week or so the pair, frustrated in their search for beef, came on an unfortunate peasant and killed and ate him.

It was plain both animals must be destroyed without delay, and a party of us went out at once to find them. There was very little difficulty about it: in fact the pair succumbed with almost disappointing docility. There was no more cattle raiding after this and it was obvious that it had all been the work of this couple alone. The other surviving lions had taken to living off game. It is probable that they moved away considerable distances to the east and south, ranging beyond the frontiers of Gwalior, for lions were several times reported from both directions and at least one was subsequently shot by Maharajah Panna to the east. A long time later I heard that the late Maharajah Baria had killed an African lion far to the south by the Kunoo river in Madhya Bharat.

Recently there have been suggestions that our native Gir lions should be re-introduced into other parts of India. The idea is a good one provided suitable sites are chosen, where the lions can do no harm to domestic stock and the jungle is free from tigers. Rajasthan is one of the provinces in which appropriately isolated, semi-desert zones can be found, and perhaps it is not quite absurd to hope that our wild lands will one day once again see these brave and handsome beasts. Certainly some such scheme is their only hope of survival: their present range is far too restricted for the race to perpetuate itself there indefinitely; there are no more than some four hundred individuals and a sudden murrain or other unforeseen mishap could wipe out the breed forever.

Though the Gir lions are of course preserved it is sometimes considered necessary to keep down the population, and a few years ago I went with the Maharajah of Jaipur on a shooting party organized by the Jam Sahib of Nawanagar during which three males were killed. The system used was to drive the beasts as on a big tiger shoot.

Like the two African man-eaters we had to destroy in Gwalior these lions were less interesting and exciting to hunt than tigers. I was surprised to notice how little use they made of the available cover. On each occasion the lion strolled out casually into the open, offering an almost un-missable shot. My impression is that these lions are not only astonishingly bold, but, as it were, scornful of taking cover in the face of danger. Wounded, they behaved with the same boldness and ferocity as a wounded tiger.

There are many differences between the habits of the two animals. Tackling large and dangerous game a lion is more likely to 'fight loose' and strike out with his forepaws, a tiger to use his talons to grip his victim while he gets the necessary purchase with his jaws. Lions again tend to live a more social life, combining sometimes in quite large family groups and hunting in teams. A tiger, unless it be a tigress with cubs, or a male interested in a tigress—who may still have partly grown cubs with her—hunts alone. The pride, or group of perhaps a dozen adult or semi-adult lions, has no equivalent among tigers.

When a tiger and tigress are together, the tiger will attack the quarry first and make the kill. Only after her mate has eaten all he wants and left the carcass may the tigress and any cubs she has with her start feeding. When a tigress kills it shows she is hunting alone or with cubs. If a tigress brings down an animal in which her mate is interested he will punish her with teeth and claws for her presumption. A lion on the other hand, like a panther, will make way for his mate and if she kills first may allow her first turn at the feast.

As a normal rule I should expect a lion to be more open and bold than a tiger, and not half so sagacious. There is no doubt that the tiger is much the stronger and more dangerous animal.

* * *

SPEARING A TIGER IN ASSAM

Tigers are shot and hunted by various methods; I have already described over a dozen methods of killing them but the method by which we hunted a tiger in January 1958 was unique and unparalleled.

It is said that in ancient times some of the Kings of Assam used to surround a tiger with high nets and then disturb him with a few staunch elephants. The tiger after breaking cover would invariably attack the men standing outside the net. These men were armed with spears which they lunged at the tiger through the net.

The Columbia Broadcasting Unit became interested in taking a movie of this sport, and volunteered to pay all expenses if it could be arranged. As a result preparations were soon started, and a net measuring four hundred feet in length and eight feet in height with six-inch mesh was made out of half-inch thick jute rope. Five hundred spear-men were also enlisted, each with a ten-foot-long spear made of strong bamboo stick.

The buffalo baits were tied in good patches of jungle where they were soon killed by the tigers. We had about twenty-six elephants by the help of which the tigers were to be driven to a convenient spot where they could be ringed round, though in many attempts this object was not achieved as the tigers succeeded in escaping into the thickly wooded forest.

After these futile attempts we finally succeeded in achieving our object on the 28th January. On this occasion the tiger had killed a buffalo a few hundred yards north of the Rawta Range Office, where the jungle consisted of thick and tall elephant-cum-tiger grass. It was about a thousand yards between the Range Office in the south to the thick tree forest in the north. On the east was a forest road running parallel to this patch. On the west was the Rawta river where the grass jungle was rather thin and some places along the river bed were completely bare. We knew the natural line of retreat of the tiger; he would either go towards the north into the thick forest or across the road to the east. There were a few tall semal trees in this grassy jungle and we took precaution to post several lookouts in those trees to give us the information about the movement of the tiger. The jungle formed a bottleneck where it joined the thick forest in the north, and our intention was to drive the tiger towards that patch where he could be easily ringed round in a net.

Tigers generally do not walk straight in a beat. Their habit is to get away to the flanks. In order to prevent them from escaping like this we put white sheets of cloth along the road on the east and along the river bed on the west. These sheets of cloth were two hundred yards apart opposite the point where the beat was started, but narrowed down to fifty yards apart near the bottleneck, and here the net was put across between the sheets to prevent the tiger from escaping into the thick forest.

The tiger was lying close to the kill. About ten o'clock in the morning, as soon as the line of elephants was formed, we advanced in a line taking care to make no noise. Nothing could be seen on account of the dense tall grass. Only the movement of the grass indicated the position of the animal. The tiger first tried to cross the road to the east but when he saw the white sheet in that direction he stopped and turned about towards the river bed on the west. He had not gone far when he came up against white cloth again. Having now discovered white sheets on both the flanks and elephants to the rear he started moving towards the bottle-neck where fortunately in addition to the thick grass there were a few bushes and trees affording ideal places of concealment.

The tiger stopped among these bushes to hide, but the lookout up one of the trees pointed out the exact spot where he was lying. We therefore halted and surrounded the place with men and elephants. I was sure that finding himself surrounded on every side the tiger would not break cover unless he was forced to do so.

Our next operation of surrounding the tiger in a ring of net started at once and took us nearly an hour. The net was hung from a line of bipods each of which consisted of two strong bamboo posts set on the ground four feet apart. The bamboo posts were tied to pegs to prevent them from slipping, and the net which was hung vertically from the apex of the bipods was also secured with pegs to prevent the animal escaping under the net. It is a strange habit of wild animals that they do not try to jump over a net. In this case I guessed that the tiger on seeing the human beings standing close to the net would attempt to attack them through the net, thereby giving an excellent opportunity to the men to spear him.

When the ring was completed we started throwing stones into the thicket where the tiger was hiding, but the animal did not move. A few twelve-bore gun shots were fired into the thicket but even that did not disturb the animal. These operations went on for nearly an hour but had no effect. Ultimately we decided to dislodge the animal with a Land-Rover station-wagon which had a closed body. Mr. Borris and Mr. Bob Young volunteered to enter the area to force the animal to break cover. The Land-Rover did a very good job of trampling the grass and eventually forced the tiger out. As soon as the tiger got up he saw the line of human beings standing behind the net barring his way from entering the thick forest. At once he charged them with a roar, but was soon pulled up sharp by the net. Several spear thrusts from the men outside convinced him that there was no escape that way so he quickly returned to his original cover.

There the Land-Rover started worrying him again, and this time he attacked the people who were obstructing his way of escape into the jungle on the south—the direction from where he had come. Again he received a hot reception by the spear-men and so for the second time he returned to the cover. Now his left eye was seriously injured. He refused to leave the cover for a long time until the Land-Rover almost trod on his tail.

This time the tiger made a final attempt to escape towards the river bed, but he still had to face the spears. By this time the morale of the spear-men was high, but the tiger was severely demoralized. He made his final charge over a bit of open ground which gave me an excellent opportunity to take photographs. After the third charge it was clear there was no more fight left in him. In fact he had become so weak that he could not regain cover but stretched out in the open gasping. There we had to put a bullet into him to save him from further agony. In his body we found nine spear holes but the one over his eye seemed to be the vital one.

Although in my long experience tigers have been killed with swords and sticks such occasions were purely accidental and not premeditated like the event described above, when proper arrangements were made and the tiger was deliberately speared. It was not only exciting but dangerous from the beginning to the end.

K. Ullas Karanth

J.D. SCOTT

FORESTS OF THE NIGHT

The cat walked slowly, almost indolently, along the ribbon of yellow dirt, a road that bordered the tiny Indian forest village, sauntering as if on the way to one of the thatch-roofed huts to report in for supper and a nap. Mary Lou, sitting beside me in the jeep, gulped, and Rao cut the motor and gently brought the machine to a silent stop. Dusk was just beginning to seep into the jungles of the central province of Madhya Pradesh, a dusk that arrives like a sneak thief, and vanishes before you are aware of it. Then night comes down as if someone had yanked a curtain. The brazen cat was a leopard that had already entered the village of Bori and killed a calf tethered just outside its owner's hut. It had dragged the calf away, hidden it, and now was back to lord it over the unarmed villagers. Its assured, graceful stroll was a frightening thing to watch.

We were patrolling the perimeter of the village because the natives were terrified of the animal that seemed to have no fear of man, and Rao was afraid that the leopard might happen across one of the villagers. Then we would have real tragedy on our bands.

Rao just pointed his finger at Mary Lou. She looked steadily at him for a moment, slid across me, got out of the jeep and, with shooting vision already beginning to reach its dim and dangerous stage, she checked the .308 rifle, carefully following the walking leopard. The rifle barrel moved as steadily as if it were on a stand. When the sound came, it was a sharp, sudden crack that startled us even though we sat waiting for it. Almost instantly the spotted cat crumpled, kicked both hind legs, then was still. Rao, the driver and the forest guard immediately started talking exuberantly in Hindustani, stood seemingly at attention for a moment, then bowed their heads slightly, in respect to Mary Lou's accuracy and calm. She wasn't watching them; her rifle was still at her shoulder and she was completely focussed on the cat in the road, unaware of the noise of the four Indians, of the jeep, me, everything except the fact that she had a leopard on the jungle road and wasn't sure that it was dead.

With rifle jutting before her like a huge, pointing finger, she advanced toward the terrible, suddenly immense, figure of the leopard. Rao ran forward. 'It is all right,' he said. 'He is dead! See?' He stooped, picked up a stone and tossed it at the sprawled leopard, striking it squarely in the midriff. It didn't move a hair.

Now it was dark and Mary Lou stood still a moment longer, watching the cat, then she brought the rifle from her shoulder and smiled. She was a lovely brunette who always seemed to have a golden glow about her, and she stood slim and elegant in her khaki trousers and bush jacket, looking for an incredible instant there in the wild night like a Bergdorf Goodman model showing a prospective customer how shooting clothes are supposed to fit. She was my wife, Maria Louisa, the name shortened by school chums to Mary Lou, but, for a moment, as I looked at her proudly, the whole thing seemed unreal, like something rolling off TV or romantically day-dreamed.

It was real enough, we both realized as we stood on the veranda of our dak bungalow two hours later, watching the entire village dancing and cavorting before us, paying homage for the killing of one of the most terrible creatures in their jungles, the cunning leopard, or panther as they called it.

We had come to these dark jungles on a sort of mission: I am a professional writer and much of my work seems to appeal to sportsmen and has appeared in several national magazines and in a thrice-weekly column in the New York *Herald Tribune*. One day, early in 1957, the Government of India Tourist Office suggested that it might be interesting for me to journey to India and discover by first-hand experience what sports that country had to offer. They also said that it was the first time this offer had been made.

Ashoka Dutt, a slender, handsome, young Indian with velvet

eyes, and a quick wit, was then publicity officer in the Government of India Tourist Office in New York. His understanding about visas, his planning of our itinerary, and his thoughtful, almost fatherly advice got us off on the right foot with his country and its people. He arranged to have Tourist Office representatives meet us at every stop in India, and gave us a list of what he thought we would need, even though his office didn't know too much about hunting and *shikar.* Shikar, which is Hindi for big-game hunt, is the same to India as the safari is to Africa. By working out the climatic conditions for February and March, the time of year that we would be in the jungle areas, we discovered that we would need both cold- and warm-weather clothing. We were grateful to Dutt many times late at night, as we shivered on a rough *machan* high in a palas tree, even though we were heavily bundled in alpaca duck-shooting coats and caps and lined shooting pants. Left to our own resources we probably would have taken only tropical gear into the jungles. Nearly everything we had read on India concentrated on its steaming climate. Writers like Jim Corbett never paid much attention to comfort. He believed in telling only the bare facts of the hunt and the killing of cats. He probably figured every damn fool knows about clothing and similar items. Or that no one was simple-minded enough to go after the big cats anyway.

We consulted with Henry Hunter, public relations chief of Olin Mathieson Chemical Corporation, which numbers Winchester Arms among its companies, and went to the factory at New Haven to experiment, shooting various high-calibre rifles before we made our selection. This was terribly important; our lives and perhaps the lives of others could depend upon our choice. Hunter has a keen knowledge of people as well as guns, and he was helpful with detailed information on what Winchester rifles had accomplished in the field, both in Africa and India, although the Indian information was quite sparse.

Believing that one should stay with rifles with which he is familiar, and having respect for and a working knowledge of Winchester's bolt-action Model 70, we tested that model in the .458, America's most powerful big-game rifle, a whale of a weapon that hurls a .510-grain bullet. Technically the rifle has a muzzle velocity of 2,215 feet per second, with a muzzle energy of 5,110 foot pounds. We read documented reports of the rifle's knocking down an elephant with one shot. We didn't want to shoot an elephant and never have been able to understand why anyone should, but we did want to get the dangerous great gaur, some of them weighing in excess of 17 cwt. We also wanted enough gun for the tiger. The .458 might be a bit too much for the big cat, but I have always believed in over-gunning rather than under-gunning.

For lesser game, like the antelope, leopard, wild boar, blue bull, chital and sambar, we selected the reliable and powerful .308, also in Model 70. Its muzzle velocity, with the 150-grain silvertip controlled expanding bullet, is 2,860 foot seconds; the corresponding muzzle energy, 2,730 foot pounds.

We also took along my favourite, a Browning superposed twenty-gauge 'Lightning' shotgun and a Winchester Model 21 twenty-gauge for dove and the famous Indian grey partridge. I brought several boxes of the three-inch No. 6 cartridges for water fowl.

I found that the Scandinavian Airlines System (Air India International wasn't flying into New York) knew a lot about sportsmen and their transportation problems, having trail-blazed the polar routes and instituted the polar-bear flights, so I talked with George Herz, their public relations manager. One of the basic reasons for the success of our trip was meeting and knowing people like Ashoka Dutt, Henry Hunter and George Herz, men who knew their jobs so well that they made ours easy.

Armed with a Government of India permit to take the guns and ammunition into their country, and a customs record of export, so we could bring the guns back into this country, we had little trouble getting them aboard the plane. There is on most planes an ammunition limit of 22 kilos per aeroplane compartment. That's about 48 pounds, and as we followed airline instructions and had all of our cartridges in the original wooden box, the whole business weighing about 40 pounds, we were all right on that item.

We learned later from Allwyn Cooper, Wardha Road, Nagpur, our shikar outfitters, that we could have contacted them and shipped guns, ammo and heavy clothing on ahead, say a month early, thus saving ourselves the trouble of lugging the stuff with us. I'll probably do this if I tackle a shikar a second time. However, having had them misrouted on a trip to Mexico, I had learned that it is wise to carry your guns with you. Our rifles, bolts off, were in heavy sheepskin cases, and the shotguns in flat leather cases. We carried our cameras and binoculars.

We discovered that a vaccination certificate is required for a person who arrives within nine days of his departure from or passage through a yellow-fever endemic area. The Passport Division of the State Department gave us health cards which were filled out and signed by our doctor, John Street, who also, with the help of the Slone Brothers of New Milford, put together a kit of medicines, antibiotics, the latest and most effective malaria pills and the like. We were required to have smallpox, typhus and cholera inoculations. Dr Street also advised typhoid, paratyphoid and tetanus shots.

We put off a couple of inoculations until the night before we departed. We were to be in Dr Street's office by eight o'clock, but when we started out through nearby Bridgewater, Connecticut, on a snowy night, a house at the road junction was totally ablaze and fire engines and cars blocked us for about an hour. When we got to the office, the doctor was out on an emergency call; consequently we didn't get home until well after midnight. We were well punctured, but felt protected. The needles hadn't let our enthusiasm leak out. At least not mine.

But by this time, Mary Lou was beginning to think that maybe this trek to India was not such a good idea. It took me two hours to convince her that she had a form of stage fright and that everything was going to be all right.

'All that time up there in that plane,' she said soberly. 'I don't like it. I don't think I'll go.'

Shaming her by snide references to John Foster Dulles travelling one hundred thousand miles in a year changed her mind, and we started packing at two-thirty a.m. for a trip that would last months. In bed at four, up to six, and then there we were in Idlewild, to find that our SAS flight was five hours late. Finally, though, we were in Scandinavian Airline's big DC-7C, winging our way to Calcutta.

Fifty-three comfortable hours in the air, plus stopovers at Copenhagen, Dusseldorf, Geneva, Rome, Cairo, Karachi, totalling no more than six hours and SAS had us in India.

If you've never been confused by a city, I suggest that you try Calcutta, known even in India as 'the city of confusion'. The population, swollen by homeless refugees from Pakistan, is so vast that it makes New York City (even though it is larger) look like the suburbs of Cleveland. Your first impression as you stand at Dum Dum Airport is that somehow you personally have kicked over a fantastic human ant-hill and the swarm is upon you: six coolies—in soiled red, dark blue and white turbans and dhotis that even at some long ago point were a dubious white—appear for each of your bags, and jabber and gesticulate. A great tug of war starts, with children anywhere from six to twelve grabbing your coat, your sleeve, your lapel; asking—with smiles, eloquent brown eyes and palms outstretched for rupees—whether you want a taxi, a guide. They are not obnoxious about it, and you get the feeling that they have this privilege in this land where poverty is a way of life to be fought with any means.

The courtly Colonel M.D. Framjee, head of the local tourist office, arrives. With his soft, assured English and the help of his two skilful assistants, K.K. Roy and K.C. Chakravarty, the confusion is gone, the customs cleared, the bags stowed and we are in a black Studebaker, brushing bullock carts, goats, making split-second, nerve-fracturing stops to avoid hitting oblivious people, chickens and sacred cattle (sacred because they were supposed to have been created on the same day as Brahmins). Through streets that teem with people—sitting, walking, sleeping on the pavements, preparing their meal of a wheat mash along the kerbs—we make our way to the Grand Hotel, a gigantic, implausible pile full of equally implausible characters: smug, smiling, well-dressed Japanese businessmen; a few harassed-looking English; Ghurkas, Sikhs, Punjabis, Marathas, Madrasis; Russians in their bell-bottomed trousers and with bovine expressions; a few unquiet Americans, conspicuous in this off-tourist season. A page ripped out of an E. Phillips Oppenheim novel.

One day, one hearty mutton curry and one savoury chicken curry later, we were back at Dum Dum Airport catching the midnight mail plane to Nagpur, the city known as the geometric centre of India, and our jumping-off point for shikar.

Once you've checked through customs (usually an uncomplicated and friendly procedure if you are a tourist), there is little trouble with baggage. Overweight is checked and paid for. Then the coolies take over, quietly, effectively. The Indians, as I was to realize many times, are a people with a highly developed anticipatory intelligence. They know what should be done, what you want, before you do.

Overnight I discovered that I was getting a reaction to my typhus inoculation and my left arm was inflamed, swollen and painful. The plane contained only three-seat units and Mary Lou and I were separated: she jammed between two brightly sari'd women; I with a handsome but grumbling Anglo-Indian on one side and on the other a huge, silent, bearded Sikh, who smelled slightly of garlic.

As I glanced at the glossy beard of the Sikh, the Anglo-Indian (half Indian, half English, of a minority group that seems to be held in contempt by both races at this point in history) immediately went into detail on what the Sikhs were, much to my embarrassment and that of the big, bearded Indian. Nearly everyone in India understands English and most speak it. It seemed to me more of a national language than Hindustani.

'These chaps were pushed about in a bloody fashion by the Moslems a long time ago,' the Anglo-Indian said. 'They decided to give themselves that fierce look by growing beards and all that rot. Protective colouring sort of y'know.'

He was quiet for a moment, then staring directly at the Sikh, went on: 'Their religion could be called the five Ks, if you like. *Kes*, the unshorn hair, held by the *khanga* comb, covered by the usually bright turban, the *kara* metal wristlet, the *kacch* underpants, and the *kirpan*, kind of a sword. Unlike most Indians, they can take a drink when they want to. And most of them want to.'

On his lap the Anglo-Indian had a sleek calfskin mail bag with a brass lock. His mission in life he told me was to deliver the mail to an official at the airport at Nagpur, then return to Calcutta, carrying another full bag. The bag, from the intense way he clutched it, must have been jammed with money-mail or registered letters. He had English features, with soft, brown Indian eyes—an interesting combination. As he chatted with me, he idly polished the brass lock with the sleeve of his tweed jacket. It gleamed like a diamond.

As he bumped against my arm that was ballooned and getting very sore, he told me that Nagpur was a good place to stop 'for a half-hour or so, chap. Excellent fresh orange juice at the port. But no place for a chap to sleep. Bugs, y'know.'

I had developed a violent dislike for him and his habit of banging my arm every time he wanted to talk, and by the time the plane sat down I was hoping that something would happen to cancel his return flight so he would have to bed his pompous little self down with the bugs at Nagpur.

As we deplaned, a tall, darkly handsome, slender man, wearing horn-rimmed glasses, and dressed in the high-necked coat, the *achkan* of the Indian diplomat, stood on the tarmac just outside the entrance. He had been watching us from the moment we got out of the plane, and although there were two other American couples and two Englishmen with their wives, he walked toward us, and said, 'Mr and Mrs Scott?'

When I said yes, he shook hands with me, and then placing his hands together before his face, prayer-like, in the graceful Hindu *namashthe* greeting, he half bowed to Mary Lou.

'How did——' I started.

'You have the look of a hunter,' he explained smilingly. The ability to use astute flattery, intermingled with tact and good manners to the point where it emerges as sincerity, is another Indian talent. This was Vidya Shukla, owner-director of Allwyn Cooper (Private) Limited, the shikar outfitters who were to handle our hunt. A member of the House of Parliament, a bright young Indian who knows his way around, Shukla will be heard from in Indian politics if he isn't knocked off by a tiger. Big-game hunting, especially cats, is one of his favourite hobbies—the reason for Allwyn Cooper, where he can mix business and pleasure.

As we sat waiting for our dozen bags to be gathered, we made conversation with Vidya Shukla. 'We're looking forward to shikar,' I said a little fatuously. 'I've been wondering how it differs from African safari——'

He gave me a quick, inquiring glance. 'It differs all right,' be said seriously. 'I think that they are a bit better organized, get more co-operation from their government than we do. But I also think they are more commercial. I believe I can give you a quote, a rather long one if you don't mind, that might answer your question.' He hesitated. 'Have you ever beard of Captain C. Forsythe?'

'Is that the man who wrote *Highlands of Central India*?' Mary Lou asked, getting an appreciative nod from Shukla and a surprised expression from me. She had spent hours in the New York Public Library before we left and had gleaned some amazing facts.

Then, quoting freely and brilliantly, Shukla said, '"I will here speak only of the glorious field that India offers to the sportsman—incomparably the finest in the world. As a field for sportsmen Africa may be thought to be better, but it is not so if India be looked at as a whole. Perhaps, more animals in number or in size may be slaughtered in Central Africa; but that does not surely imply superior sport. In reading accounts of African shooting, I have often wondered how men could continue to wade through the sickening details of daily massacre of half-tame animals offering themselves to the rifle on its vast open plains. In India, fewer animals will perhaps be bagged; all have to be worked for, and some perhaps fought for.

'"The sport will be far superior, and the sportsmen will return from India with a collection of trophies which Africa cannot match. Africa and India both have their elephants. We cannot offer a hippopotamus; but we have a rhinoceros superior in a sporting point of view to his African relative. We have a wild buffalo as savage and with better horns than Cape species; and we have four other species of wild bovine besides, to which there is nothing comparable in Africa. In felines, besides a lion, a panther, and a hunting-leopard, almost identical with those of Africa, we have the tiger, and one, if not two, other species of leopard. Our black antelope is unsurpassed by any of the many antelopes of Africa; and besides him we have fourteen species of antelopes and wild goats and sheep in our hills and plains, affording the finest stalking in the world.

'"Africa has no deer at all, except the Barbary stag which is out of the regular beat of sportsmen. India, on the other hand, has nine species of antlered deer. We have three bears; Africa has none. There is no country in the world that can show such a list of large game as we can in India. And for minor sport, what can compare with our endless array of pheasants, partridges, dove, pea fowl, jungle fowl and water fowl?"'

'That's what we call quite a mouthful,' Mary Lou said. 'How can you remember it?'

Again that warm smile. 'Many of my clients ask me the same thing: How does shikar compare with safari? The English hunter, Forsythe, knew his subject so well and wrote about it so eloquently, I thought it only a mark of respect to use his words.'

I had never been entertained before while waiting through the dull routine of having bags sorted at an airport. 'This is the best-timed example of soft sell I have encountered,' I said. 'Madison Avenue could learn something from you.'

'I am happy that this gets us off on the right foot,' he said. 'I believe that successful and pleasant shikar is more important in making friends for India than millions of printed and shouted words coming from Parliament, and I have told Nehru so.'

He drove us to the Mount Hotel, a veranda'd, one-storey tropical hostelry right out of an old Warner Brothers' movie. 'I'd like you to breakfast with me,' Shukla said. 'I'm leaving tomorrow afternoon before you take off and I'd like to say good-bye and good luck over coffee.' He bowed and departed. We were alone in the middle of India.

By now, although I hadn't mentioned it, Mary Lou noticed my arm and was terribly concerned. She tried to make the room bearer, a little, wizened brown man who appeared as if from a hole in the floor, understand that she wanted some ice. Failing to get her point across, she rolled up my sleeve, showed him the swelling and touched his hand to the blaze.

It was now that still hour of the morning, about three-thirty, an hour when everything seems strange and exaggerated even when you are in familiar surroundings, and the two of us felt not only exhausted but completely isolated, cut off from the rest of the civilized world. A rat scurrying in the far corner of our room didn't enhance this first night alone in India. But I pointed out to Mary Lou that we had bedded down in similar circumstances in remote Canada and Mexico, and that by daylight we would have our perspective back.

The bearer padded back in twenty minutes with some ice cubes and a couple of bottles of mineral water. Two things that we had been repeatedly warned about by old Indian hands like our friend, world-traveller Fred Rosco, were to skip all raw vegetables and to drink no water unless it was bottled or boiled.

The Mount Hotel room bearer also had a note for us: 'Dear Scotts: We're on our way home and would like to see you in the morning.' It was signed Dan and Marjorie Maddox, Nashville. Tennessee. It was the buoyant note we needed.

The whole thing had the familiar Hollywood touch: the Humphrey Bogart Hotel, the message from white hunters just out of the jungle, me with an arm that felt as if it was about to split open; an Indian who spoke little English, padding back and forth in a sleek, white turban and dhoti; the brunette (Mary Lou), who kept ice packs on my burning arm for hours; the beds with their gauze or mosquito-netted canopies; the large fan turning lazily and hopelessly overhead; the open spigot in the next room that served as bath, shower, water supply and tormentor as it dripped continuously.

Daylight did come, however, and the arm deflated. Coffee and tiger talk with Shukla made the day right, and he told us that he would send S.V. Rege, Allwyn Cooper's officer-in-charge, to travel by train with us to Harda, the edge-of-wilderness town where we would be met by Rao Naidu, our *shikari* or hunter. Mary Lou and I were sorry to see Shukla go; he would have been fun and a great comfort to have in the jungle.

And then Dan and Marjorie: Dan was tall, slender, dark, and articulate; Marjorie, blonde, lovely and lively. They had gentle Southern accents and more enthusiasm than I have ever encountered. At first the talk was all jungle, all tiger. Then it got into the supposedly mundane things like food and clothing and weapons. But nothing is mundane with the Maddoxes.

'My dear girl,' Marjorie said to Mary Lou, 'y'all are just going to freeze if you park yourself on one of those machans. You goin' to do it?'

Mary Lou looked at me. 'Yes,' she said.

Marjorie went to one of her bags and rummaged, coming up with some long-john underwear. 'You'll need these,' she said to Mary Lou. 'And these'—giving her a pair of canvas slippers—'for around camp.'

We then had some tea, made robust with Dan's priceless bourbon, and more talk.

'You're in for a surprise,' Marjorie said to Mary Lou. 'Club 21, the Carlton House, La Crémaillére à la Campagne should do as well. You'll be staying in the same dak bungalow we had, with the same staff. The cook, Motisingh, is so darn yummy that I hated to leave——'

'Blue bull,' Dan broke in. 'Ask for some of that loin of blue bull that I left. It is the best thing I ever ate——'

'Nope, suh,' said Marjorie. 'Peacock! That's the greatest; simply the most delicious hunk of stuff in the world!'

After ransacking their bags again, they gave Mary Lou and me sweaters and heavy caps, telling us that the nights were bitter cold and we'd need them sitting up for tiger. Dan even offered to lend me his .458. I had one and he was relieved; he thought it was the rifle for anything big and dangerous that walked in the jungle.

'I took a trophy of nearly everything worth-while,' he said quietly. 'There are two more tigers there in our blocks, I know, and a great bison that I missed, that is a prize animal, a giant. I would have sworn that I hit him, but the shikari said no. That's Rao Naidu, quite a guy. Knows his stuff and you'll like him——'

'Dan got his tiger and leopard and missed a beautiful tigress,' Marjorie said. 'I think he respects the leopard more than any other animal, and he's been to Africa and all over the place, hunting.'

'A testimonial,' Dan said. 'But she's right. I got more of a bang out of that leopard than anything, including the tiger. They're smart as hell, and really dangerous. They've learned to live closer to man than any other wild animal, and they know our habits better than we do ourselves. There are literally thousands of them and they've gotten to the point where they enter villages, and carry off cattle, kill people.'

We had to meet Mr Rege, get our liquor allotment—which Maddox told us was quite an undertaking—check our clothing and guns and get ready to board the afternoon train. The Maddoxes said they would see us off.

'Mr Rege, Shir S.V. Rege, was the officer in charge of Allwyn Cooper, which meant that he was the man who took most of the responsibility of getting clients into the jungle and did most of the work at home base. He was short and soft spoken, a gentle version of Edward G. Robinson, and he lugged a bulging, well-worn leather brief-case that immediately gave the impression that here was one son-of-a-gun of a busy man. The impression was correct: he bustled like a Boy Scout leader with his first troop and shooed us around Nagpur getting our liquor permits signed, checking on our personal belongings, reminding us of items we should have.

'Kleenex,' he said pensively, several times. 'Don't forget this Kleenex. It comes in handy in the jungle.' He was right.

It was as if Nagpur were a giant drum with the population jumping up and down on top of it, yelling at the top of their lungs: the lowing and bleating of cattle and goats in the streets and on what served as pavements; the constant blare of motor-car horns, which seemed to merge into one long, consistent sound, never stilled, always shrill and irritating, as if horn buttons of one hundred cars had stuck simultaneously. And the colour: turbans and saris of every hue; brilliant flashes of red, gold, purple, fine yellows and pale blues that stood out sharply among the brown and white bullocks and the black water buffaloes; the squalid dirt roads, the run-down shops, lifting the whole medieval scene into one of fine art, limned for ever in your mind.

That is the lasting impression of India: the incredible number of people and the fantastic colour everywhere.

Our last stop was at a government office where we spent forty minutes with a high clerk (who wanted to be pompous but didn't know how), filling out long impressive-looking forms, attesting (we think) that we weren't going to become alcoholics or sell liquor to the natives. Then we went around to the government shop and got twelve bottles of Indian beer (excellent!), two of English gin (horrible!) and one bottle of rye whisky, with a brand name we couldn't decipher (no comment!), and that was it. Rege was amazed that we didn't take our full quota. It wouldn't have been much more than we had, but he kept saying, 'Americans always take their quota.'

Back at the Mount Hotel the Maddoxes were waiting to go to the station with us, and Dan, strangely enough for a big-game hunter, spent ten minutes effusing on the wonder of Nagpur oranges and how he didn't know how he was going to survive in the future without them. 'You'll see,' he said darkly. 'They drug you with them. Then when they're gone, your life isn't worth living.'

Our baggage had been sent to the station ahead of us and was there in a great heap, interspersed with wicker and woven-bamboo baskets that were filled with comestibles like eggs, oranges, bananas, and surrounded by a widening circle of staring Indians. This was India's Central Railway and the antiquated train was waiting, shaking as if it were going to come apart at the rivets. Rege had reserved a huge compartment for us, with a stand-up type lavatory and section for baggage. On the way, we had made one stop at the Ashoka Restaurant and Rege had bustled from it with a wicker hamper covered with a spotless white linen napkin.

Dan bent over one of the baskets, and brought out a handful of small oranges, gave us each one and we stood there in that incredible railway station, eating the sweetest oranges I have ever tasted, talking about when we would see one another again, and surrounded by what seemed to be half the population of Nagpur.

Then Rege bustled us aboard, got in with us, opened the windows of the car, and Mary Lou and I stuck our heads out, shouting good-byes to Dan and Marjorie Maddox, who stood eating oranges and waving. The entire crowd at the station made noises like a college-football cheering section as we rattled out of Nagpur.

PART 2

PRESERVATION AND NEW NATURAL HISTORY

A. HOOGERWERF

JAVAN TIGER

Sody (1933, 1949[a])[1] has more or less clearly demonstrated that there are differences between the tiger populations living in Sumatra, Java and Bali, to which in fact at a much earlier date the names *Panthera tigris sumatrae* (Pocock), *P.t. sondaica* (Temminck) and *P.t. balica* (Schwarz) were given. Although at first differences in size were presumed between the representatives of these different populations, this later (Sody, 1949[a]) proved to be untenable, except for the Bali tiger, which is supposed to be smaller than those of the other two islands. The principal difference between the Sumatran and the Javan tiger is the darker basic colour of the latter's coat, while specimens from Bali are darkest. Although Schwarz in his original description of the Bali tiger expressly stated that the ground colour of the coat is somewhat brighter and the light markings clearer, Sody (1933), who had the opportunity of investigating the extensive collection of the Ledeboer brothers, representing tigers from Sumatra, Java and Bali, concluded that specimens from the latter island are darkest. There is also said to be some noticeable disparity in the shade of the inner front legs, which also does not agree with the findings of Sody when compared with material from Java.

According to A.M. Husson (in litt.) the classification given above is in general still regarded as the correct one today.

In Indonesia, tigers occur only in the islands mentioned above. The assertion that the species has never lived in Bali (Meiszner, 1957) is incorrect, since material from this island may be found in the Zoological Museum, Bogor, and the big-game hunters A.J.M. Ledeboer, A.F. Wehlburg and D.G. Wolterbeek Muller, among others, shot tigers there. According to accounts by Zimmermann (1938), a planter and tiger hunter who spent many years in Bali, at least 14 tigers were killed there in the period from 1933 to 1937, of which seven by him. In the Hunting Regulations for Bali, 'Peraturan Pemburuan Bali 1949', the tiger is placed in the category of 'harmful game'; a tax of Rp. 50 was levied on hunting of this predator.

Doubtless the tiger was regarded as a serious threat at the time the very first Dutchmen arrived in Java, as can be seen from the great attention devoted to it in the old chronicles. On many occasions the depredations of tigers were listed in the '*Daghregister gehouden in 't Casteel Batavia*', the official Government diary of those days. The harm done was so extensive that a bounty was placed on the head of every tiger killed, this premium being increased on September 1, 1747, to 10 crowns—an unprecedentedly high amount for those days—presumably because the lower reward between mid-October 1746 and the end of August 1747 had yielded 'only' 26 tigers. This higher amount resulted in no less than 80 tigers being killed in the vicinity of Batavia (Djakarta) between September 1, 1747, and January 14, 1749, demonstrating the numerousness of tigers in those days. However, in 1762 the premium was completely withdrawn.

But the trouble caused by these predators was still considerable at the beginning of the nineteenth century (Anonymous, 1866); the nuisance attracted so much attention that in the *Bataviaasche Courant* of February 26, 1820, it was suggested that a company should be set up to exterminate tigers. In the second half of the previous century, too, the harm done was apparently still very considerable: in 1862, 148 persons and in 1863, 131 persons in Java were killed by these predators, which led to the bounty system being restored in 1862; this time £30 was paid for each tiger killed.

Although superstition often restrains the Javanese from killing tigers if they have done no harm, in Tegal alone £2,000 to £3,000 a year was paid in bounty! In 1862 reference was even made in the Second Chamber of the States-General of the Netherlands to the large number of victims of 'the ferocious beasts' in the Netherlands Indies.

Not until 1897 was the premium abolished; however, in 1966 the NITOUR tourist agency—illegally—reintroduced a reward amounting to US $25 for every tiger killed in order to encourage hunting by tourists from abroad!

As late as 1907 Kaledjetan kampong, east of Udjung Kulon, had to be evacuated on account of a tiger plague (Anonymous, 1932). Kal (1910) states something similar with regard to some villages inside Udjung Kulon, which had been rebuilt after the eruption of Mt. Krakatau in 1883. Since the beginning of this century, when these animals were not uncommon in sparsely populated areas, the stock in Java has declined rapidly, so that there need be no doubt about the great rarity of this species in this island, driven back into a handful of localities, of which Udjung Kulon may be regarded as the principal pied à terre. It is even a sound assumption that the Javan tiger, together with the Javan rhino, is on the point of being exterminated. The big-game hunter A.J.M. Ledeboer endeavoured for many years (1910–1940) to kill as many tigers as possible, for which purpose he had at his disposal a network of informants all over Java. R.H.A. van Maarseveen (in litt.) shot two tigers in South Garut in 1927 and four (a male, a female, and two juveniles) in 1933 in the Tjisadea-Tjilaki region (South Tjiandjur); in the same region a poacher shot another tiger in that year, whilst probably a sixth tiger was hit there by a set-gun. As far as the present author has been able to ascertain, this was the largest number of tigers established in the last 50 years in Java within such a relatively small area in so brief a period of time. In the 1930s reports were still circulating fairly regularly in the press and in periodicals about the occurrence of tigers in Java. By the beginning of the 1940s the population of this predator proved to be very low, for an inquiry held by Sody in 1940 among the members of .the Netherlands Indies Hunters' Association resulted in only two reports for that year; there was none at all for 1941 (Sody, 1941[d]). In October 1938 a tiger was caught in a gin-trap near Tamandjaja, a few kilometres east of the Udjung Kulon reserve. In July 1938 a mare and her foal were killed by a tiger near the Tjikudjang rubber plantation (about 30 km northeast of Udjung Kulon) and in December of that year two people (Pfanstiehl, 1939). In 1939 a tiger was shot in South Banjuwangi (East Java) on a young banteng bull which it had killed (Anonymous, 1939[a]). At the beginning of August 1940 a dead tiger was found in the Subang area (Eastern West Java) and a second specimen was seen there in October (Veldman 1940). In mid-June 1940 a tiger was killed at a place where this species had no longer occurred in living memory, viz. near Tjibadak between Bogor and Sukabumi, West Java (Bartels, 1941[b]). Most probably two tigers were killed in South Malang in 1940. M. Bartels shot another two in 1940 and 1941 in South Bantam; at that time a third specimen was reported there (Bartels l.c.).

The extremely small number of reports outside the reserves which it was possible to trace since 1930 and which have been given above make it probable that tigers have now become a rarity in Java.

* * *

THE SITUATION IN UDJUNG KULON

Besides the following details almost exclusively relating to the sanctuary, many of the particulars published in Chapter 33 were also obtained within this area.

Older Accounts

One of the first serious visitors to Udjung Kulon, F. Junghuhn, mentioned these predators along the eastern boundary of the present reserve around 1850. He wrote that some twenty of his Javanese bearers drove off a tiger in the night of 14 May, 1846, and then seized the latter's prey, a giant turtle still alive, which it was an effort for six of them to carry.

After a trip on foot through the area the botanist S.H. Koorders recorded many deer and tigers.

As already stated above, in 1907 the Kaledjetan kampong situated just east of the sanctuary had to be entirely evacuated on account of a tiger plague; something similar seems to have happened with the villages of Tjibunar and Pulau Peutjang (Kal 1910). According to reports from the population, this was also the fate of a village near Tdj. Tereleng.

More detailed information on this species may be found in the reports written at the beginning of the 1930's by the retired Head of the Civil Administration of Bantam, J.S. de Kanter, on visits which he paid to the area, for the purpose of shooting tigers there (Chapter 29). On each of the seven trips made to Udjung Kulon in 1932–1934 this hunter's party was able to establish the presence of one or more tigers at tethered prey, tied-up baits or otherwise. As a result of all these expeditions, in which the hunters did not scruple to use gin-traps and set-guns, only one tiger was obtained, however, four shots missing. An attempt made in 1935 by a French party, likewise equipped with a legal permit to bag a tiger, had no success at all.

Permission for these hunting parties was granted on the strength of a letter addressed in July 1932 to the Director of the Department of Agriculture, Industry and Trade—who was then in charge of affairs relating to nature protection and wildlife management—in which de Kanter (then still Head of the Civil Administration of Bantam) drew attention to a surplus of tigers which, he said, had not failed to have a highly adverse effect on the stock of banteng and other game living in Udjung Kulon. De Kanter went so far as to advise that the tigers be thinned out by means of poison or that rhino, banteng and deer be transferred to Pulau Panaitan. At variance with this presumed scarcity of game, however, was the fact that during one single visit de Kanter shot three muntjac for tiger bait; this was done without legal permission. In fact there was every reason to doubt the value of de Kanter's recommendations, since he was very keen to shoot one or more tigers.

Other visitors to Udjung Kulon at the beginning of the 1930s also reported the presence of tigers. During the many visits paid by the present author to the reserve in the period from 1937 to 1957, many tracks were found and a tiger was seen no less than nine times, reaching a climax from 1937 to 1941; in the postwar years he never set eyes on another one, though spoor were rather regularly encountered and during 1955 the guards reported having seen four tigers. The daily reports drawn up by these people revealed that in 1939/40 once two tigers and on another occasion even three were seen together; a tiger was encountered 13 times in 1943, including on one occasion (9 January) a tigress with a cub and also once (7 April) a pair of these predators. Towards the end of 1940 a banteng cow, evidently killed by tigers, was found on two occasions, shortly after one another, and two months before a banteng bull. In the preceding months many tracks and faeces had been noted. The guards, too, complained in those years about the boldness of these animals.

No trip was made by the present author to this sanctuary without tigers betraying their presence by spoor, droppings or otherwise. But the symptoms were the most pronounced during the prewar period mentioned above, as demonstrated previously in discussing their effect on the banteng population. And yet the author never acquired a satisfactory picture of the overall tiger population permanently living or periodically present in Udjung Kulon, and therefore never proposed measures for checking their number. Suggestions in such a direction were also dropped in view of the very vulnerable position of this species in Java and because of the possibility of a natural check caused by some disease in connection with the discovery of three dead specimens (p. 262).

The author never gained the impression that there were too many tigers in Udjung Kulon or that predator pressure on any game species inside this reserve was excessive. The number of banteng killed was probably not inconsiderable (Chapter 29), and yet during all the years in which the present author visited the area the stock of these animals was satisfactory to good, in fact perhaps exceeding the carrying capacity of the sanctuary in the prevailing conditions; moreover, there were indications that among the banteng which fell victim to tigers were individuals weakened by distomatosis and malnutrition. Wild boar and barking deer always remained numerous and deer seemed to increase in number. Nor did the lack of observations by the author himself of tigers running in pairs and of spoor of a tigress with a cub or of very young animals going alone justify the assumption that this species was too numerous here.

The well-known connoisseur of tigers in India, F.W. Champion (1934: 57), is of the opinion that the species living in normal circumstances does not increase to such an extent that the supply of animals forming its prey becomes insufficient. He writes: 'The number of cubs normally seen with a tigress is two or three, although it would appear that some of the cubs in the bigger litters are eaten by the parents—presumably to prevent a possible shortage of food as a result of too many tigers in a limited area.' And further: 'He (the Hon. J.W. Best in the *Indian Shikar Notes*) states that, in the Central Provinces, the jungle tribes universally believe that the male parent eats the spare cubs and even records an actual case, at Bilaspur, of three quarter-grown cubs having been found partly eaten by a tiger.' Elsewhere (Champion 1934[a]: 75): 'At first sight it would appear that the carnivora, having no enemies, must breed so rapidly and increase to such an extent as to eat out their own food supply, and thereby destroy their own means of existence. Yet in actual fact this does not happen, as has been observed many times by explorers penetrating into wild and lonely parts of Africa. These explorers have found places totally uninhabited by man where the relative numbers of lions and ungulates always seem to be correctly adjusted, and it is difficult to explain how this is done. Probably Nature arranges that the lions breed more slowly when they suffer no casualties, and certainly the larger and more powerful males drive others away from their hunting grounds as soon as they find that there are too many of their tribe in the neighbourhood.' It was partly these pronouncements that contributed to the author's policy of those days.

Cannibalism, as found elsewhere in carnivorous animals, also occurs among the tigers living in Indonesia. In Bali Ledeboer observed how a tiger shot in the afternoon of 20 August, 1910, was dragged away by a fellow-tiger only an hour later and was later found partly eaten (Ledeboer 1941[a]). In Sumatra a tiger shot in the evening was on several occasions found the next morning largely eaten by another tiger (Muller 1941; Jacquet 1941[a]).

True, these cases relate to dead tigers, but there is no reason to suppose that the situation here would be any different from that on the mainland of Asia or that a threatening surplus of tigers would not be eliminated in natural conditions in areas like Udjung Kulon.

The author has never regretted not having suggested that tigers be killed off, despite the great opposition, especially from game hunting circles. Nevertheless, now that Udjung Kulon is predestined to become the last stronghold of this species, the possibility of an unnatural situation seems in no way excluded, so that the development of the population of this and other predators should be closely watched.

Encounters with Tigers

The author saw tigers on nine different occasions, of which six times in full daylight, almost without exception in the late afternoon. All these visual observations took place in the years 1937 to 1941 and none in the postwar years. Although some of these encounters have been put on record before (Hoogerwerf 1938[d], 1939[a], 1947 and 1947[a]) a brief account will be given here.

On 11 September 1937—after a lengthy period of drought—a tiger, probably a female, was seen in the midst of the Tjigenter river—containing clear water—in heavily wooded surroundings. The animal was observed for a brief moment around noon from a distance of about 50 metres, before it disappeared into the heavy forest. It is practically certain that the tiger was taking a bath.

On 3 May, 1938, a tiger (probably also a female) appeared on a rainy afternoon from dense undergrowth fringing the Djaman swamp pasture. This specimen had probably been induced to leave its cover as a result of a number of alert banteng cows giving vent to an alarming snort, announcing the tiger's whereabouts even before the author could see it. For the rest the tiger paid little if any attention to the alarming banteng, but all the same disappeared soon afterwards in the scrub jungle close to the tree in which the author was sitting to observe banteng.

A third encounter took place on 28 October, 1938, also in the middle of the day, during a dry, hot period. This time it was a male tiger which came down to the Tjikarang river to drink at a place which in those days was much visited by banteng. The photos on plates 60 and 61 were taken on that occasion. Although the animal clearly reacted to the sound of the camera shutter, he appeared to withdraw normally thereafter into the surrounding light forest.

In the afternoon of 30 October, 1938, a tiger was twice observed, perhaps a female, in the Niur area; probably this was the same specimen on each occasion. In this case, too, the animal's presence was announced some considerable time before it appeared from the thick undergrowth at about 3.30 p.m. However, whilst the author was readying his camera, the animal disappeared again. About 6 p.m. a tiger (the same one?) again appeared about 200 metres away from this spot in the open field—an extensive area covered with mainly *Cyperus* sp. and *Phyla nodiflora*—where at that moment a small herd of banteng and a flock of deer were grazing; the tiger apparently did not notice these at first. Not until it had walked some 100 metres in the low vegetation, which was about 40 cm high, was the tiger spotted by these deer; only then was their loud cry of alarm heard, and they 'packed' closely together, staring as 'frozen' at their foe. Now the banteng were alarmed too; however, these did not stand still but headed towards the predator, forming an irregular semi-circular chain, with the calves closely behind the cows. Although the tiger was at first walking upright, it later (perhaps first made aware of the deer by their cries of alarm) stalked along in a crouching attitude, now and again holding its tail almost vertically. Until it was about 50 metres away from the deer the tiger maintained this crouching posture, but then, apparently after having noticed that any further stalk would be useless, reverted to a normal one and—as if the deer did not exist—made for the banteng, looking as if it meant to by-pass these animals. Finally it passed within five metres of some cows without vouchsafing them a glance. However, the banteng did not allow themselves to be passed on the flank, but jumped aside in such a way that they were facing the passing predator again. After this the latter slowed its pace, licked its right shoulder a few times and walked straight for the spot where the author was in a blind built in a high tree. The deer remained standing in the same place, but the cows followed their foe at a distance of 50 to 75 metres. After having been briefly invisible behind a large block of coral by a muddy waterhole, the tiger made for the water, lay on its belly and lapped it up from a shallow pit in which much of its head remained hidden. The cows, still loudly alarming, were probably the reason why 'stripes' repeatedly raised its head. Then also the warning screech of peafowl was heard, mingled with the clamour of kingfishers and other birds, and the tiger's walk began to resemble a shameful retreat. For the first time it now also started roaring, which was repeated at regular intervals, and it seemed as if the animal was making a wide circle around the waterhole. The tiger's roar was regularly answered by the alarming bellows and snorts of banteng.

In the late afternoon of 3 January, 1939, it was probably once again the alarming snort of a banteng, this time of a bull, that on the Tjidaon pasture led to a male tiger emerging from a thick vegetation of old alang-alang and shrubs, walking straight across a small open patch to disappear again in the bushes without he or the bull launching an attack. The bull did not even lower his head to intimidate the tiger.

About four months later, likewise in the late afternoon, a tiger was seen along the edge of the Tjigenter pasture; this time the animal had probably not been forced to leave cover, but was later spotted by a few grazing banteng, giving vent to loud alarm signals. With head erect and tail raised almost vertically, the predator strolled between the sparse trees around the open pasture and disappeared in the direction of the heavy forest, intensely watched by the banteng, which, however, did not follow him. Shortly afterwards there was much roaring by a tiger, probably from this animal,

On 17 October, 1939, the seventh encounter took place, again with a male. This observation was made from a hide constructed in a tree by a waterhole during a moonlit night in the Niur area. The animal left the thin undergrowth at 8 p.m. and lay down on the dry, bare ground just in front of the author's blind not more than four metres above the ground; like a cat before the fire he lay licking his paws. After some hesitation the camera was focussed on the recumbent animal and two flashlight photographs were taken, which the tiger apparently ignored. The animal probably became disquieted by an alarming banteng bull standing head up, in a very alert posture, a few dozen metres away, apparently wanting to drink at that waterhole. The tiger rose and approached the drinking place, from which he drank, from a tree trunk lying in the mud. Then he walked, not far from the bull, which snorted loudly but did not move, across an open plain to the forest opposite, without the banteng following and without the tiger appearing to take any notice of him.

On 16 October, 1940, a freshly killed banteng cow was encountered not far from the mouth of the Pemageran. Much of the carcass, which clearly displayed claw or teeth wounds on the neck and throat (Plate 64), had already been devoured. The remains were tied to a tree nearby with well-camouflaged rattan; above it a blind was built into which the author moved that same afternoon, watching until the next morning. At about 9.30 p.m. the leap of a tiger was heard, after which the carcass was pulled loose and dragged into the surrounding bushes as if the body were not tied up at all. Until about 10 p.m. the sound of crushing bones was heard almost continuously, and the next morning little remained besides a few bones and parts of the cow's hide, as is shown on Plate 65. The author did not see more than a flash of the tiger as it bounded at the carcass, breaking the thick rattan as if it were a piece of string.

The ninth and last encounter took place on 21 February, 1941, in dark weather at about noon—the first and sole observation in the south-west monsoon—when a tiger jumped away from a half-eaten banteng bull in a thick lamiding (*Acrostichum aureum*) vegetation. This was along one of the Njewäan meadows, and it was one of the few times that the author has felt himself threatened by this species, although there was perhaps no reason for any fright.

In the night of 3 March, 1953, a large male tiger passed within a few metres of the author's hide along the south coast beach near the mouth of the Tjitadahan. Since the blind was closed at night on account of the bad weather and the wild breakers on the beach suppressed any sound, nothing was noticed of this strange passer-by;

the next morning the spoor showed that the predator had been about 5 metres from the blind, which was built a few metres above the ground, and had sat on his hindquarters close by. Strolling alternately in and beside the *Pandanus* vegetation, this tiger had travelled almost 2 km along the dunes and the adjacent beach.

As stated above tigers were also repeatedly observed by the guards, especially in the prewar period, but also repeatedly after the war. On 14 July, 1954, at 11.30 a.m., one was seen by the Tjitelang and on 21 August, at about noon, another one near Tdj. Tembing; in both cases they were probably males. In 1955 a tiger was seen on 7 January on the patrol path near the Pemageran at 2 p.m., on 4 February at 7 a.m. near the mouth of the Tjihandeuleum, on 9 July at 9 a.m. on the patrol path between the Tjidaon and the Tjibunar, and on 27 July at 11 a.m. near Kaledjetan along the eastern boundary of the reserve. In all these cases the animals observed were said to be tigresses. On 6 March, 1956, two tigers, apparently of opposite sex, were seen near Tdj. Tjikaret along the extreme west coast of the reserve and on 26 July, 1946, a banteng bull killed by a tiger was found at the Tjigenter pasture.

A remarkable incident is said to have occurred at Karang Randjang on 2 May, 1942. At 6 p.m. on that day some guards were eating their evening meal together with their bearers, sitting on the ground of the Karang Randjang bivouac before the open door, when a tiger leaped inside and flung itself at the bivouac coolie Masjdji, who was the furthest away from that door. It grabbed its victim, ran through the bivouac with him and finally bounded away in the direction of the beach. The next morning only the head and a thighbone of the man were found in the surrounding forest. As far as is known, this was the only occasion that a human being fell victim to a tiger within this sanctuary.

The following story told by Guggisberg (1966) relating to Udjung Kulon is untrue: 'It is said that after the Second World War a gang of poachers entered the reserve in order to slaughter all the rhino for their horns. The men beat a hasty retreat after one of their number got himself killed by a tiger.'

PRESERVATION OF THE UDJUNG KULON TIGER

Notes on the preservation of the Udjung Kulon tiger may appear irrelevant when considering the communications by Schenkel (1967, 1969) and others (Anonymous, 1968) that this sanctuary no longer harbours a single tiger. However, the author's experiences in Udjung Kulon with this predator justify the inclusion of this section.

It is extremely difficult to make specific suggestions on the preservation of the tiger in this area, on account of the many questions which had to remain unanswered, relating to the number of tigers living permanently or periodically in the reserve and their effect on the fauna, in the first place on big game, and further with respect to reproduction, about which hardly any data at all could be obtained.

Inside Udjung Kulon no species of animal appears to be threatened with extermination by these predators or to have reached an inadmissibly low level because of predator pressure, with the possible exception of the rhino now that it may be assumed that the annual killing of one or two baby rhino could mean the end for this species.

If there is in fact proof that *Rhinoceros* calves are killed by tigers, there is no reason to assume that this would not happen in Udjung Kulon. As long as indications of tigers having killed rhino have not been found, no measures may be proposed for artificially reducing the tiger population, since this predator is likewise a species on the verge of extinction. Apart from the above it has been found that the stock of tigers living in Udjung Kulon can hardly be influenced, in view of the great difficulty in getting tigers to visit laid-out bait or in bringing them within man's reach in some other way.

Should it ever be established that the tiger population of Udjung Kulon were reaching a dangerous minimum, transmigration from other parts of Java would no longer be possible today, whilst introducing tigers from Sumatra would be irresponsible, since it is assumed that the two populations do not belong to the same subspecies.

The present author does not know on what data A.F. Köhler (in *Survival Service Commission Red Data Book, I.U.C.N.*, 1966) based his remark: 'The survival of the Javan tiger is dependent on the effective administration of the Udjung Kulon Reserve but because of degradation of habitat the Reserve lacks adequate numbers of prey species and the area is not suitable for long-term conservation of the tiger.' The second part of this comment is completely incomprehensible, because there are several animal species on which tigers prey that find an excellent habitat in this reserve. Moreover, the author fails to see what might be the shortcomings of Udjung Kulon, covering about 115 square miles, making this area unsuitable for these predators.

Although opinions may differ about the role that tigers play inside a certain area with regard to natural selection, it may be assumed that in regions like Udjung Kulon they have an important function in maintaining a natural equilibrium and a favourable physical condition of their prey.

Preservation of the Javan tiger must stand to the fore, but we must endeavour to find out more about its effect on the life of *Rhinoceros sondaicus*, whose survival is still more important and more urgent.

Editor's Note

1. *This article contains references to sources and photographic plates which have not been reproduced in this collection. However, these references are in the book* Udjung Kulon: The Land of the Last Javan Rhinoceros *by Hoogerwerf. Many of the sources listed are archival material in Dutch.*

Praveen Bhargav

E.P. GEE

TIGERS

Apart from the Corbett National Park in Uttar Pradesh, Kanha must be one of the best places in India for seeing tigers.

The tiger is believed to have entered India from northern Asia after the last ice age. It must have come through China, past the eastern end of the Himalayas range and into north-east India. It could not reach Ceylon, which was separated by sea. It has been recorded in all parts of India except the desert regions of the north-west and the higher Himalayas.

The present-day Manchurian or Ussuri tiger is a larger beast with a longer coat. It is to be presumed that the slightly smaller and thinner coated tiger of India is a modification evolved over the years due to the greater heat and moisture of its new habitat and possibly due to the fact that smaller prey has to be killed in India than in the pre-history days of northern Asia.

The term 'Royal Bengal' which is sometimes added to the tiger is fanciful and not recognized by zoologists. There is only one subspecies in India, and the tigers found in Bengal are not necessarily larger or more handsome than those in Assam, Bihar or other parts of India. I don't know how the word 'Royal' originated, unless it was from a tiger shot by the Duke of Windsor when he was Prince of Wales; but the word 'Bengal' could have been rightly applied, as the type specimen, from which Linnæus gave the name *Felis tigris* in 1758, came from Bengal. The tiger has since been renamed *Panthera tigris*, incidentally.

Tigers are usually solitary in habit, except at the time of courting and mating and except in the case of a tigress with large cubs. A tiger's 'territory' is not a compact area of country but is generally in the form of a 'beat' or sequence of places visited at regular intervals in search of prey. Should another tiger from outside venture on to such a 'beat', there might be a fight—unless the weaker one gave way without a struggle.

There has been a lot of controversy about the tiger's power of scent. Some affirm that it has feeble powers of scent, while others quote instances to prove the contrary. Probably the real truth is that its power of scent is good but that it does not use it or need to use it as a rule, for its powers of sight and hearing are so remarkable.

Tigers used to be very common in India, but improved and greater numbers of firearms, electric torches and spotlights of jeeps and so on have reduced their numbers to only a fraction of what they were. Also the fact that a tiger constitutes one of the best of all sporting trophies has contributed to its decline. I don't suppose there are more than about 4,000 tigers left in the whole of India to-day, contrasted with a possible 40,000 of fifty years ago.[1] In some parts of India, such as Madras, they are becoming quite rare, though in a few other places they are still fairly abundant.

Due to the great decrease in the amount of natural food in the jungle, deer and pig, for instance, the few remaining tigers and leopards are often forced to come out and prey on domestic stock, thus becoming 'cattle-lifters'. Forest-dwelling tigers are leaner due to the more active life they lead, while cattle-raiding tigers become very bulky due to the easily found prey on which they are feeding.

Very rarely a tiger will either accidentally or purposely kill a man—but will not eat him. But once such a man-killer tastes human flesh, it may acquire a liking for it and become a man-eater.

Most man-eaters, however, become so due to some wound or other disability, or to old age, which makes the pursuing and killing of a wild animal, or even a domestic one, rather difficult. They evidently find killing a human being a very easy matter. The cubs of a man-eating tigress would naturally grow up to be man-eaters themselves due to the acquired habit and taste.

A man-eater is very rare—only about three or four in a thousand.

You are quite safe in a forest and the normal tiger will not harm you—unless you accidentally intrude on a tigress with young cubs or a wounded animal, or possibly an animal fast asleep.

I remember once walking for a mile or more through some tall grass. What was presumably a cow was walking away just in front of me—I could hear it and could see the grass moving as the animal walked ahead. Eventually I came to an open place, and there was the 'cow'—a huge tiger.

During my last stay at Kanha I was going out in the jeep one afternoon about 4 p.m. with the Range Officer when we met four men returning in an excited state. They had seen a tiger on a kill. We took them on board and went straight to the spot.

We left the jeep and walked about half a mile. Some vultures were in the trees—sure sign of a kill. Soon I, who was leading, caught a glimpse of a tiger looking up angrily at being disturbed, and then making off along a fringe of tall grass into the forest. A dead bull gaur was lying there half eaten, probably killed the previous night.

No wonder we had failed to see any gaur that morning on our trip round the area. I felt sorry for the fine dead beast, but this is the law of the jungle, the rule of predator and prey, and there are usually surplus bulls in a herd of gaur.

It was now 4.30 p.m. with about two hours before it became completely dark. The driver of the jeep suggested that news be sent back to the Conservator of Forests who was at the Rest House, an idea with which I did not like to disagree.

So I decided on a plan of action. I would sit on the ground behind a diminutive bush some distance away from the kill, and the Range Officer could stay with me if he wished to. There was a good chance that, after the driver and the other four men had noisily departed and the jeep moved off, the tiger would return to the kill. Photography was out of the question, for the setting sun would be directly into the lens of my camera: it would have to be a case of 'tiger watching' only.

I have occasionally sat *up* (in a tree) over a kill for a tiger in my younger days, and have experienced the unforgettable thrill when the magnificent creature arrives. Here was a chance to sit *down* near a kill, and experience the same thrills without the shooting.

I told the Range Officer of my plan, and asked him if he would like to stay with me. 'Ye-e-s' came the reply from a man obviously a bit nervous about the shape of things to come.

Now I should here reiterate that a tiger, unless molested by man, is one of the safest and most gentlemanly creatures of the jungle. There is no fury on earth greater than that of a tigress with small cubs if she is disturbed in her hideout. A wounded tiger of either sex is positively dangerous at any time and in any place. A man-killer, or worse still a man-eater, is an uncanny and deadly peril and the news gets round quickly if one is at large.

A solitary wild buffalo or elephant can be unpredictably dangerous at any time. Even a rhino or a boar can be fearsome. But this looked like a straightforward case of a tiger or tigress at a kill. I have never been less afraid.

I selected a tiny bush about three feet high and got the Range Officer to sit behind it. I sat more in the open because I knew I could remain absolutely still. The bush would conceal the movements which I knew from experience my companion would make. The other men departed, and soon we heard the jeep start up and move off.

After a while, some of the vultures flapped down to the ground, near the kill.

The Range Officer looked longingly up at a tree nearby, up which we would have been very safe but very uncomfortable and in full view of the tiger. I beckoned to him to remain seated where he was. His continuous movements, fearfully glancing all round the place and wiping off the perspiration, made me think that perhaps the tiger would see him and not come out. Luckily the small bush hid his movements and nervousness. The sun started to sink, just above the kill.

A movement at the edge of the forest. Something was cautiously moving round, to get a clearer view of the place. Yes, it was the tiger . . .

After staring straight at me for some time, it strode majestically across an intervening patch of open ground into the long grass near the kill. Then it, or rather she because it was a tigress, was joined by another slightly smaller tiger. It looked like a tigress and a nearly fully-grown cub. They came up to the kill. The mother started to drag it farther away, but stopped after a while. They both moved about—quite unaware of our presence.

Then they decided they were not so awfully hungry, and could wait. Had they not eaten more than half a gaur? And had they not come back quickly with the main object of driving away those impudent vultures? So they sat down in different places and cleaned themselves like huge domestic cats.

A grand sight indeed! And no shot fired, no ear-splitting report, no shattering of the silence of this peaceful scene, no wounded or killed animals. Nothing to disturb the natural spectacle of India's most magnificent predators going about their lawful business.

It was getting late, so we left our tiny bush, and the tigers quickly retreated into cover. The Conservator had come and we told him all about our experience. As it was soon to be dark we all returned to the Rest House.

I asked the Range Officer if he had been afraid during the episode, and he readily replied, 'Yes, very much.' 'Why?' 'Because I have not had such experience of tigers.' Which goes to confirm that many people hold this mistaken notion that all tigers are dangerous.

But what a marvellous story he told the others later on in the evening! How brave we were, and how close we were—the seventy yards or so of safety between us and the tigers had somehow suddenly dwindled to seven! Anyway, it was good to know that he, like myself, thought the affair very much worthwhile.

I envied him his gifts of description and exaggeration. I myself suffer from the failing of dry factuality and lack of imagination. As I heard him retail the episode to his friends I was reminded of a story I once heard from a friend of mine: A group of village *shikaries* (sportsmen) were swapping yarns round a camp fire about tiger shooting. One old man, not to be outdone, enthralled the others by recounting how he had once fought a huge tiger and eventually killed it with a large knife. It measured eighteen feet nine inches.

The others protested at this impossible measurement, and continued their protests. The teller of the story then magnanimously conceded that it all happened a very long time ago and that he *might* be mistaken. And that the tiger *might* have really been nine feet eighteen inches!

* * *

Also in the State of Madhya Pradesh but some distance away to the north-west, is the national park of Shivpuri. This place, like several other parks and sanctuaries of India, used to be the game preserve of a ruling prince. When Independence came to India in 1947 and princely states became merged with the new States of the Indian Union, the protection given to Shivpuri by the former Maharajas of Gwalior suddenly ceased, and excessive poaching took place.

During this time most of the wild life, sambar, nilgai or blue bull (an antelope), chital, chinkara and others were unfortunately killed off. When I first visited the place in May 1957 the wild life was starting to re-establish itself. The park superintendent told me how he had carefully driven the neighbouring forests with scores of beaters, to coax back scattered remnants of wild life into the new park area of some sixty-one square miles.

That he had been fairly successful in restoring the numbers of wild life was proved by the fact that each day I saw a few chinkara, those dainty and elegant chestnut-coloured gazelles, as well as chital, sambar and nilgai. Peafowl were everywhere, and a large variety of birds including paradise flycatchers. Once while we (about five of us) were standing in some open scrub country discussing various things, I noticed a purple sunbird hen fly up, hover and then proceed with building her nest *only a few feet away* from us. I asked everyone to remain exactly where he was, and I slipped away to fetch my cameras. Result—several ciné and still pictures of this bird very close, building its nest.

The countryside and forests round Shivpuri are quite different from those of Kanha. The trees are mostly deciduous, and from March to June is 'summer' and the place is all dried up—except the fine artificial lake, and one or two perennial streams. 'Spring' comes with the monsoon in July, and autumn, as expected, in the closing months of the year. Parts of the forests are rocky and barren, but it is very good tiger country.

It was in this area that several record tigers are reported to have been shot. But I am personally always rather sceptical about measurements of tigers (and leopards), because there are many opportunities, for those who wish to, to enlarge these measurements. The correct way to measure the length of a dead animal is, of course, to lay it flat on the ground, drive a peg in the ground at its nose and another at the tip of its tail, remove the animal and then measure the distance between the pegs (wooden, not whisky!). In some parts of India, I am told, they measure round the curves of the animal; and the more they press the tape into the softer parts, the greater will be the length of the tiger.

In the old days it used to be the custom that a very important person, say a Viceroy, must shoot a really large tiger. So when the VIP was not looking (I give him the benefit of the doubt in this!) the organizers of the shoot would stretch the tiger as much as possible and push the tape as much as they could into all the soft and hollow places in order to produce a large tiger. I have been told by one famous sportsman that a special tape, with eleven inches to a foot, was used for Viceroys!

Not far from Shivpuri is a stone, recording the 'record' tiger, shot by Lord Hardinge in 1916, measuring eleven feet six and a half inches. In 1924, slightly farther away, a tiger measuring eleven feet five and a half inches was shot by Lord Reading. I wonder . . .

There is a sort of viewing lodge at Shivpuri, where the Maharaja and his guests could sit under cover and watch a tiger at a kill. A 1000-candle-power electric arc lamp was hung overhead, with a special switch to regulate the light from five candle-power to full, so that photographs could be taken. After the tiger had made its kill, the light was gradually increased from 5 to 1000 candle-power in a time of ten to fifteen minutes, so that the tiger did not notice the oncoming 'daylight'.

I did not use this place myself. But I heard the story of a certain Forest Officer who had obtained permission to use it. A tigress killed at 3 a.m., and the light was gradually increased from 5 to 1000 candle-power. The tigress then withdrew, and later returned with four small cubs. They all started to feed.

Then a huge male tiger appeared, unexpectedly.

The tigress hastily led away her cubs, concealed them in the nearby jungle, and then returned to deal with the intruding tiger. As the tiger moved around the kill the tigress alternately placed herself between the tiger and her cubs, and between the tiger and the kill. She kept tactfully edging the tiger both away from the direction of her cubs and also from the kill. This went on for some time. The tiger, deciding that discretion was the better part of valour, withdrew—though it was evidently hungry.

Then the tigress returned to the nearby thicket and brought back her cubs to their dinner. A wonderful drama enacted just a few feet in front of the Forest Officer—but I did not hear if he got any photographs of it all.

The park superintendent of Shivpuri, Vijay Singh, seemed to me to be most knowledgeable about wild life, and genuinely interested in the subject. We had long and absorbing discussions. He told me how the tigers had been arranged for Marshal Tito a short time previously, and how such arrangements are usually made.

If a full month's notice is given, arrangements can be made for at least one tiger to be at a given spot at the exact time required, even in daytime. It is done by tying live buffalo 'baits', at regular intervals which vary according to size of tiger and whether one or two tigers are coming.

I must now be careful not to get involved in the old controversy as to whether live 'baits' should be allowed or not for tigers, lions and leopards. I myself do not approve of it, but neither do I strongly disapprove of it: I am inclined to accept it as a custom that has been going on for thousands of years. To some it may sound cruel—but I would remind such people that when the matter has been examined from every angle there are pros as well as cons, and a balanced view needs to be adopted. An anthropomorphistic viewpoint that the tied-up animal must endure agonies of suspense and fear while it awaits its fate, unable to escape, is incorrect.

The arm-chair reader living in a city may perhaps be invited to reflect that a tiger has a right to live and that to do so it must kill some live animal about twice a week; that wild herbivorous animals in the forests have become scarce, so a tiger will come out and kill domestic stock anyway; that the tiger is a very selective creature, and will kill the healthier and younger animals and overlook the useless and aged if it has the chance to do so; that there is an alarming surplus of every kind of domestic stock in India, from buffaloes down to goats, which are consuming all the vegetation and doing their level best to turn the country into a desert as many parts of North Africa and the Middle East have become; that when live 'baits' are required only useless young buffaloes are purchased from the owners, who naturally do not part with their more valuable animals; that the live 'bait', when tied up, usually grazes or complacently chews the cud oblivious of the swift and skilful death it faces; that the alternative—a long walk along hot and dusty roads to a slaughterhouse where the killing may be delayed a day or two—would entail much more suffering to the animal.

I have mentioned all these points because if I had avoided the issue altogether I would have laid myself open to the criticism that I had shirked it!

The gratifying thing that I heard from Vijay Singh was that Marshal Tito declined the offer of shooting a tiger, but instead he commendably took ciné films of it. Or rather of them, because no less than four tigers turned up for him and his party to see all in one spot, and not one of these tigers was shot—except with cameras.

Although that great sportsman and naturalist Jim Corbett did not do very much wild life photography himself, and although he did not travel much in other parts of India, yet in his *Man-Eaters of Kumaon* he comes down strongly on the side of the person who 'shoots' with a camera rather than with a rifle.

He wrote '. . . the taking of a good photograph gives far more pleasure to the sportsman than the acquisition of a trophy; and further, while the photograph is of interest to all lovers of wild life, the trophy is only of interest to the individual who acquired it'.

I remember reading in a newspaper report that sometime in the early forties Jim Corbett told the Viceroy, Lord Wavell, that the tiger in India might become extinct within ten years.

The tiger has not become extinct, though twenty years have elapsed since Jim Corbett's prediction. Thanks to its cunning and preference for thick habitat, it has been able to survive in India, though its numbers are becoming lower as each year passes by. I think that eventually, due to increasing human population, it will only be found in zoos, and in some of the larger national parks and sanctuaries—provided that State Governments can fully protect these fine wild places.

It must not be presumed from the foregoing that wild life conservationists in general, or I myself in particular, object to the shooting of tigers in India nowadays. Tigers can multiply quickly in certain localities, and a tigress rearing two or three cubs at regular intervals can produce a temporary local tiger population which, in an inhabited area, would have to be reduced by shooting.

Moreover *bona fide* sport, when the shooting rules are observed and necessary permits and licences obtained, is always encouraged by enlightened preservationists, for it brings in much needed revenue, and the presence of sportsmen in an area is invariably a deterrent against poachers. The out-and-out protectionist, who objects on principle to the killing of game animals and birds for sport, has no proper place in a progressive country, where surpluses should be harvested for revenue and wild animal populations kept within reasonable limits.

It was for this reason that I was surprised at the outcry in Britain because the Queen's party shot a tiger during the royal visit to India in the spring of 1961. An inquiry had been addressed to me several months previously from the Fauna Preservation Society in London, asking if it was ethically and conservationally all right for a tiger to be shot during this visit. I replied that the tiger is not a protected species in India, and that the shooting of a tiger by the royal party would be just the same as shooting a stag in Scotland, and that therefore it should not be frowned upon.

A VIP guest in India is still traditionally offered a tiger shoot; and provided that tigers in any particular area have not become scarce, there is no harm at all in this.

If sportsmen were not available for shooting tigers in places where they have become a menace, then professional killers would have to be employed to do the job—at great expense to the State.

On the other hand, the shooting of a cow rhinoceros in Nepal by a member of the same royal party was a tragic affair. For the rhinoceros is fully protected, even in Nepal, and this was a cow with a calf. In extenuation it should be admitted that the royal host in this case probably considered the rhino as 'royal game', and the royal guests could only have declined the shoot at the risk of being thought discourteous.

Note

1. I have stated my reason for giving rough estimates in the penultimate paragraph of page 29.

Editor's Note

It is rather amusing to note that these rough guesses of tiger numbers by Gee made without any basis are widely being repeated by conservationists and the news media forty years later!

GUY MOUNTFORT

SAVING THE TIGER

I was convinced that there was still a fair chance of saving at least the Indian race [of tiger], which the world knew best as the 'Royal Bengal Tiger', providing that three conditions were fulfilled. First, the scientific resources of the IUCN would have to be brought to bear on the techniques involved in such a difficult operation. Second, the willing cooperation of the governments concerned would have to be obtained. Third, the considerable cost of such a multi-national effort would have to be underwritten by the World Wildlife Fund.

I had taken negotiations concerning the first and third conditions as far as I could. The task now centred on the second and I therefore set out to see whether I could interest the heads of state in my proposals. I knew from previous negotiations that I had undertaken on behalf of the World Wildlife Fund, that if I could obtain the backing of the head of state, all doors would be open to me when it came to working out the details with government departments; whereas if I opened negotiations at lower levels I was often frustrated by bureaucracy.

My first visit was to Mrs Indira Gandhi, the Indian Prime Minister. With me were Charles de Haes, now Director General of the WWF, and Zafar Futehally, Vice-President of WWF India. I knew that Mrs Gandhi had inherited a deep interest in wildlife from her father Pandit Jawaharlal Nehru. She said she had seen me on television the previous night and was therefore aware of my interest in the tiger, which she regarded as a national symbol of India.

Seizing this opening, I outlined my proposals. If the Indian government were to support us and would also create a number of special reserves in areas where tigers were still relatively numerous, the IUCN would help in drafting plans for their scientific management and the necessary research. Also the WWF would raise the equivalent of a million dollars (£400,000 at that time) so that the reserves could be equipped to the highest standards. Although it was impossible to save all the tigers elsewhere in India, if these reserves were established and effective legislation against poaching and the black-market export of skins were enforced, the tiger could still be saved from extinction. De Haes, with his wisdom in negotiations of this kind, was quick to remark that such a programme could only succeed if directed and co-ordinated by the highest authority.

To my delight, the Prime Minister agreed without hesitation. 'I shall form a special committee—a Tiger Task Force,' she said, 'and it will report to me personally.'

A little stunned by the speed of her decision, I asked tentatively if I might repeat this at my press conference that evening. 'Certainly,' she replied.

India Takes the Initiative

The Tiger Task Force was appointed the next day. The chairman was Dr Karan Singh, one of India's most dynamic politicians and at that time Minister of Tourism and Civil Aviation. My old friend Kailash Sankhala was put in charge of implementing the programme, which later came to be called 'Project Tiger'. Zafar Futehally served as the link with the IUCN WWTF. Field surveys were made and a list of reserves proposed. A six-year plan, involving the expenditure of no less than £2,300,000 ($5,900,000) was approved by the government. A further budget that extended the programme until 1984 was approved later, nearly doubling the original figure. Bearing in mind India's tremendous economic and social problems, this set a magnificent example to the rest of Asia.

Before moving on to Bangladesh I attended a big press conference convened by Dr Karan Singh. In the course of a splendid speech, he said, 'During the hundred years of the British Raj, Englishmen slaughtered our tigers. Now an Englishman is leading a crusade to save them.' I thought this rather a wry comment even though justified by history. However, it was said with a smile.

In Dacca, capital of the new state of Bangladesh (formerly East Pakistan), I quickly obtained a meeting first with the new President, Justice Abu Syed Choudhury, and then with Prime Minister Sheikh Mujibur Rahman, the leader of the revolution which had succeeded with India's aid, in gaining independence from Pakistan. Everyone was still in a state of euphoria over the victory, as well as being shocked at the devastation that it had brought. I feared that I could expect little interest in the tiger.

Nevertheless, the President welcomed my plans, saying Bangladesh must take every opportunity to gain any international assistance which could restore the country to a normal condition. Moreover, the creation of wildlife reserves would help to attract tourism and foreign currency. He promised his support, adding rather sadly that the country's natural beauty was one of its few assets to have survived the war. This was a promising beginning, but I knew that real power lay with Mujibur Rahman and that without his backing I might still fail.

I met him that evening, having pushed my way through a milling throng of officials, journalists and petitioners to the ministerial offices. Press photographers were playing with a young leopard which someone had presented to the Prime Minister. The atmosphere was chaotic and very different from Mrs Gandhi's quietly efficient surroundings.

We were being served the inevitable tea when the Prime Minister burst in, greeting us with complete informality. His secretary had warned us that he was always in a hurry. In fact, while we talked, he never stopped pacing up and down, radiating vitality and nervous energy.

I outlined my proposals again, this time against constant interruptions and questions. He obviously liked them. His face lit up and from then on he did most of the talking.

'Conservation is part of my plan,' he declared. 'The destruction of forests has been terrible! But do you know what I did? Two days after becoming Prime Minister I issued a decree forbidding the felling of any more trees and the killing of any more wild animals. The tiger? Why, it's now our national emblem and on our new bank notes! Of course we shall protect it. We shall have a great national park in the Sunderbans!'

His enthusiasm was wonderful. I began to understand how such a man could have broken through the apathy of the patient Bengali peasants and led them in a war to create a new nation. He also meant what he said. Within a few years we had not a national park, but three tiger reserves in the Bangladesh Sunderbans. Like India, Bangladesh also issued special postage stamps depicting the tiger. Poor Mujibur Rahman. Little could I know that he was soon to fall to a hail of assassin's bullets. A few years later, after I had been negotiating with President Daoud of Afghanistan on behalf of the World Wildlife Fund, he too was assassinated and I began to wonder if I carried around a jinx with me.

It was nine o'clock before we left Mujibur Rahman's office. Although I was exhausted after a week of constant meetings and press conferences in India, I now had to meet the Bangladeshi press to ensure that the campaign received good coverage. I then had to summarize the meeting for the local radio station, which I did against a background of shrill chatter and honking taxi horns which probably made it unintelligible.

Early next morning I was on my way to Nepal. As the snowcapped Himalayas appeared through the aircraft window, I felt again the familiar thrill which landing at Kathmandu always gives me. John Blower, the project manager of the local United Nations wildlife conservation programme and adviser to the Nepalese government, was there to meet me; so also was George Schaller, whom I had last seen at the other end of the Himalayas, when he was studying snow leopards in Chitral. We exchanged news as we sat over coffee in the sunshine.

The situation in Nepal had changed since my last visit. Previously. I had been able to discuss conservation developments directly with HM King Mahendra. Now, alas, he had died. He had been succeeded by his son, HM King Birendra, whom I had not yet met.

I dined that night with the British Ambassador, Sir Terence O'Brien, and Sir Edmund Hillary, who was just back from the Himalayas. Although the Ambassador tried to arrange an audience for me with the new king, none was forthcoming. The king was still in official mourning and very busy with Cabinet appointments, and could not see me for at least a week. By that time I had to be back in Switzerland to present my proposals for the tiger project at a joint meeting of the IUCN and WWF. I agreed to wait as long as possible, but meanwhile briefed John Blower, who would be involved in the implementation of the proposals if they were accepted. During the next few days I visited the Royal Chitawan National Park, to find out how its tigers were progressing, and discussed with John Blower the possibility of creating two new reserves at Sukla Phanta and Karnali, both of which I knew were excellent tiger areas. There was time also for John to take me in a light aircraft to examine the Mount Langtang area, which he wanted to turn into a national park because of its extraordinary beauty. We managed to land somewhat precariously in a stony valley at an altitude of 12,400 feet near some scampering yaks, just short of the Langtang glacier. I agreed enthusiastically with John's proposal, for the peaks and lakes in the surrounding area were truly superb.

Fortunately, it was not long before HRH Prince Bernhard, at that time President of the World Wildlife Fund, visited the new King of Nepal and succeeded in obtaining agreement to all our proposals. In addition the King's brother, HRH Prince Gyanendra, took charge of the whole conservation programme, which has made great progress ever since. Today Nepal has three splendid tiger reserves at Chitawan, Karnali and Sukla Phanta, and both Langtang and the Khumbu areas surrounding Mount Everest have become spectacular national parks.

I had previously visited the small kingdom of Sikkim and satisfied myself that very few tigers were resident here. Bhutan certainly had tigers, particularly in the Manas forests, but was very difficult to enter, so I decided that both these countries would have to be tackled later. In any case, I now had to return to Switzerland for a joint meeting of IUCN and the WWF. At the meeting it was agreed that the tiger project should now be given top priority. The WWF immediately organized a world-wide fund-raising campaign, to be called 'Operation Tiger', and to be carried out by all its national organizations. The IUCN meanwhile appointed some of its most highly qualified people to liaise with their Indian, Nepalese and Bangladeshi counterparts in planning the creation and management of the tiger reserves. Some were to be entirely new; others already in existence were to be enlarged and improved.

The creation of an effective wildlife reserve is not simply a matter of erecting a fence around a protected area. The reserve has to be skilfully planned and managed, and its wildlife constantly monitored. First the site has to be accurately surveyed to ascertain its geophysical features, vegetation and water resources. The populations of its major animal species have to be determined in order to calculate the biomass in relation to the available food sources, and the maximum carrying capacity of the reserve. In the case of the tiger this means knowing the numbers and sex ratios of its prey species. If necessary the amount of land available for grazing and browsing may have to be increased if the tigers are to have sufficient food to prevent them from wandering outside the reserve. If the deer and pigs are too numerous they may have to be culled to prevent them from destroying their habitat.

Guards must be able to patrol the reserve effectively. This involves the construction of roads and guard huts. Watch towers and water for fire-fighting have to be available and wells may have to be dug. Guards have to be recruited, trained and properly equipped with uniforms, vehicles, firearms, binoculars and two-way radio sets for controlling poaching. The cooperation of the local police, forestry officials and neighbouring civic authorities has to be obtained. Sometimes, to satisfy local requirements, a properly controlled hunting area has to be established outside the reserve, where surplus game can be shot on a sustained yield basis. Finally a long-term management plan must be prepared to cover not only the reserve but the protective buffer zone around it. Here the willing cooperation of the villagers is essential.

New reserves can rarely be established without imposing some restrictions on local activity. When planning a reserve one has to give a high priority to problems involving human interests. In Malaysia and New Guinea, for example, it was found that small numbers of aboriginal people inhabited the rain forests where new wildlife reserves were to be created. As in neither country were they seriously depleting the wildlife, but were living in peaceable equilibrium with it, no attempt was made to remove them. These reserves indeed now protect both the aborigines and wildlife as part of the same ecosystem.

But particularly in Asia, villages may have to be translocated, a matter which requires very careful handling if friction is to be avoided. Both India and Nepal have shown particular skill in this difficult task and by generous treatment have re-settled numerous villages in new sites. Arrangements have to be made to exclude domestic cattle, which occur in thousands in almost all new reserve sites. Those which enter after a reserve has been created are put in a pound and have to be reclaimed on payment of a nominal fine. In some reserves villagers are allowed entry for a few days each year to harvest essential fodder, firewood, thatching, fruit or wild honey, under a controlled programme. The annual burning of grass, which is usually a dangerously uncontrolled affair, has also to be strictly managed.

The creation of a wildlife reserve involves a considerable number of jobs for villagers in road-making, building, transporting, the hiring of elephants, bullocks and boatmen and in miscellaneous labour. Moreover, unlike many enterprises run by remote city-dwellers or foreigners, the money earned continues to circulate locally, thus enriching the community.

After the reserve has started to operate, plans usually have to be made to cater for tourism. People cannot enjoy wildlife unless they can see it. But here again great care must be exercised. Uncontrolled tourism can destroy a reserve by sheer success and weight of numbers. In the United States, for example, only 350,000 people visited the already fully established national parks in 1916; by 1978 the number had risen to 45 million and restrictions had to be introduced to prevent serious disturbance and the destruction of vegetation by tourists. Tourism, however, generates local employment and is an important source of revenue that offsets the cost of managing and guarding a reserve.

When a new wildlife reserve is created full account must be taken of the social and economic needs of the human community if the reserve is to have an assured future. It must never be seen as detrimental to man. Conservation is invariably a compromise between the need to protect wildlife and the aspirations of the local human population. In Asia particularly, where the density of the population is so high and the economic problems so great, much skill and patience are needed in balancing these priorities.

K. Ullas Karanth

KAILASH SANKHALA

PROJECT TIGER

It was my privilege to be a member of the task force for setting up Project Tiger and to prepare the draft plan. The proposals were based on the information I had collected in my field studies under the Jawaharlal Nehru Fellowship. These proposals detailed the need to eliminate all the many factors contributing to the tiger's decline: habitat destruction and disturbance caused by forestry operations and cattle grazing; the loss of prey populations by poaching and by competition with domestic stock; and the killing of wildlife in general for trophies and trade.

To re-emphasize a very important point: predators like the tiger are second-stage consumers which cannot be preserved in isolation. They require primary consumers (prey), which depend on producers (vegetation), which in turn depend on converters (micro-organisms in the soil). Any plan to preserve the tiger therefore requires a thorough knowledge of the whole range of prey populations and their habitat, including the socio-economic behaviour of humans. It is important to guard against certain fancy ideas of wildlife management whereby the habitat would be manipulated to increase the wildlife population to an artificially high level.

The broad concepts of Project Tiger were clear: it would be committed to the philosophy of total environmental preservation in selected areas and nature would be allowed to play its part fully. Management would be limited to eliminating or at least minimizing human disturbance and to repairing the damage already done by man.

We began cautiously. Instead of spreading our thin resources all over the country we decided to concentrate on nine specific areas in different ecosystems. We lost no time in producing well-documented 'management plans', and Project Tiger was launched at the Corbett National Park on 1 April 1973—by a curious coincidence exactly 20 years to the day since I started my career in the Forest Service. Some people called it a 'quixotic project'; but at least their complacency helped us to work undisturbed.

For economic reasons alone it would have been impossible to preserve the whole of tigerland, as that would have meant the elimination, or at least the drastic alteration, of a number of forestry operations. That was neither necessary nor possible in a developing country like India. The criteria for selecting the nine areas as reserves were that each should be representative of a certain type of tiger habitat; that the habitat should be as undisturbed as possible; that the population of wild animals including tigers should be adequate or their development potential high; and that there should be no competing factors like copper or coal mining, the harvesting of valuable timber, or any other vital economic or human consideration. As far as possible the reserves should be in different States of the Union so that each State should be responsible for giving the tiger reserve due priority in management.

Jon Tinker, who visited India when we were finalizing our plans, wrote an article in *New Scientist* of 28 March 1974 entitled 'Can India save the tiger?' He remarked: 'While India's plans to save the tiger are not scientifically perfect, they are politically shrewd and realistic.' Probably he had relied too much on the census figures, which were only one of the factors we considered in selecting an area as a reserve, or on Professor Leyhausen's arbitrary figure of 300 as a viable population of tigers. I feel sure that with the ecological information and biological studies of the tiger presented here, and the successes we have already had, he and others will be convinced that the project is also scientifically sound, even if there is always scope for improvement.

If the habitat is basically good and is given protection the reactions are bound to be favourable. The tiger, who is only one part of it, will automatically increase. In fact this has already happened in all the reserves. The latest count in the reserves shows the following increases: 31 to 51 in Manas, 22 to 30 in Palamau, 17 to 60 in Similipal, 44 to 73 in Corbett, 14 to 22 in Ranthambhor, 43 to 54 in Kanha, 27 to 57 in Melghat, 10 to 26 in Bandipur, and 50 to 181 in the Sundarbans. (In the case of the Sundarbans the initial count could not cover the whole area.) These increases have been compared with the figures from the 1972 census; the recent figures (556) are more dependable. Although this includes 55 cubs born this year, the figures are not all the result of reproduction; some tigers came into the area and stayed on when they found it to their liking.

The nine original tiger reserves (another, Periyar in the Western Ghats, is to be added) are plotted on the map. We would have wished to add a few more to the list, for example one on the Bastar plateau—notorious for man-eaters—in Madhya Pradesh; another in Balfakram, in the evergreen forests of the Garo hills in Meghalaya; still others in Dudwa in the Terai in Uttar Pradesh. Then there are the famous foothills and Terai forests of Cooch Behar in West Bengal; and Sariska in Rajasthan. Due to limited financial resources and other practical considerations they could not be included, but slowly they too are likely to be brought under the umbrella of the Project.

India's tiger reserves are of course not as large as Wood Buffalo National Park for bison in Canada, nor the Serengeti in Tanzania or the Tsavo National Park in Kenya, famous for their lions and elephants respectively. Because of India's high density of human population there are no large tigerlands completely free from human disturbance. But still, in spite of the heavy pressure on land and forest produce, forests of 12,000 sq. km have been set aside under Project Tiger as reserves to be scientifically managed. The central funds amount to nearly Rs 4 million, and the sacrifice by way of loss of timber and other resources is substantial.

The wildlife of a country nowadays is of more than local interest. International concern was voiced over the dwindling of the tiger, the most magnificent of the world's predators, in its natural home. To save it is an international responsibility, and the call to support Project

Tiger through the World Wildlife Fund received a prompt response from all over the world. The imaginative publicity campaign, which included painting competitions for children and the selling of special stamps to collect funds, reached almost every door. Small and large donations in cash or kind poured in and the WWF collected more than one million US dollars.

Project Tiger was piloted by its Chairman, the Federal cabinet minister Dr Karan Singh, as his personal responsibility assigned to him by Mrs Gandhi during the first four years of the Project. No conservation project is ever popular to begin with because it imposes restrictions and even sacrifices, and there are no immediate economic gains. This is especially so in the case of a project which aims at total environmental preservation and where the guiding principles are: 'Do nothing and allow no one else to do anything.' My task of implementing the Project for its first three years was certainly challenging. Experts wrote to tell me why it would fail, and even many wildlife enthusiasts had doubts about its success. Some of my colleagues laughed at the idea of a campaign to stop forestry fellings and the grazing of domestic cattle, even more at the prospect of shifting villages. There was no precedent for such a programme; and perhaps their scepticism was justified.

The scheme depended for its implementation on the State governments, who were in no mood to accept controls, particularly in the harvesting of timber. But the Prime Minister's message clearly spelled out the policy, and they accepted their responsibilities. The direction was that 'Forestry practices, designed to squeeze the last rupee out of our jungles, must be radically reorientated at least within our national parks and sanctuaries, and pre-eminently in the tiger reserves. The narrow outlook of the accountant must give way to a wider vision of the recreational, educational and ecological value.' The query, 'Is it beyond our administrative or political competence to achieve this ?' put us on our toes and we accepted the challenge. In the reserves felling was stopped, cattle were removed, and villages were shifted.

I watched the people moving their belongings, and they did so not reluctantly but enthusiastically, happy to start an easier life free from the problems of wild animals destroying their crops. I asked one man, called Jagan, of Annantpura in Ranthambhor, how he felt at leaving the home of his forefathers. He replied that all their lives he and his family had spent cold winter nights scaring sambar and wild boars away, and he was looking forward to an undisturbed life in their new home. Jagan's wife gave birth to a son—her first child after twelve years of marriage—soon after. Everyone was given a house, more land than they had had and community facilities far better than they had enjoyed before. They entered into the mainstream of the life of the country.

A prophet is without honour in his own land, and I was not sure how seriously the Project would be taken up in my home state of Rajasthan. I had my vehicle boldly painted with the tiger of the Project and the panda of the WWF and sent it to Ranthambhor. The villagers, and even the district authorities, had never seen such an impressive vehicle even in the times of the maharajas. It had the desired effect, and the people were convinced of our seriousness. I was told that they thought the vehicle had been specially sent by the Government!

Having enlisted the support of the district authorities we were able to make the necessary restrictions really effective, and the reserve took shape so quickly that it was the first of the few to be free from cattle. The others were not far behind. Generous donations of speed boats for the delta of the Sundarbans, elephants for the rain forests of Manas, camels for the arid land of Ranthambhor, and bullock carts for Melghat enabled our protection forces to move much faster than would otherwise have been possible. For the use of the flying squads each reserve has been provided with jeeps with diesel engines to save on running costs. The poachers were broken, physically and psychologically, both by these methods of apprehending them and by the weight of public opinion against them. Our use of radios further demoralized them and gave confidence to our own men in the interior where means of communication were poor.

All this has paid huge dividends. The habitat has improved in all the reserves and the population of tigers has increased—it was good to know the tigers themselves seemed to approve!

One of our main problems was to recruit the right people to implement Project Tiger. A scheme of this nature needs zeal, faith and dedication. How were we to find such committed personnel? After the inevitable teething troubles we acquired a dozen hand-picked officers, and I took them to East Africa to gain experience of national parks. This visit cemented their dedication. They have become involved in the Project not for promotion nor for extra money—which is hardly there—but for the sake of their convictions and the faith they have developed for the cause of conservation. I had my satisfaction when one of them, Sinha, said that the prospect of only eight years left to work for the service was far too short, and he wished he had become involved with wildlife earlier. Before, he had hardly handled a camera in his life, but soon he is to have a one-man show of his elephant pictures explaining their ecology.

In Kanha, Panwar is mad about his Minolta camera, capturing the ecological changes in the meadows and studying the causes of depletion of the vanishing swamp deer. C.B. Singh protests even about angling with line and rod in the Ramganga River in Corbett Park; he is a purist. Debroy exchanges fire with poachers even at night in Manas and they know full well that he is a crack shot. Wesley's wife, but for her age and the responsibility of grown-up children, would have gone to the law court over his absences. Koppikar has lived most of his life as a married bachelor so his new life-style in the Project does not bother him; he prefers a bullock cart to a jeep. Fateh Singh's wife is a Rajput girl who is used to neglect as part of her traditions, so she is unlikely to protest. Saroj is virtually living with Khairi, a tigress in the Similipal hills. The commitment is total and the team is perfect.

In India we who manage the reserves are all foresters. In forestry we are used to the slow growth of trees and do not expect to enjoy the fruits of our labour in our lifetime. But the success of Project Tiger in just two years has dazzled even the field directors. No axe falls on any tree, no saw moves on dead or fallen wood. Dead animals are left on the forest floor once again to become a part of nature with the help of scavengers and termites and bacterial converters. Even bleached bones and shed antlers are not allowed to be removed for what some people call 'better economic use' because even dried meat fibre is somebody's food, and bones a source of calcium. Death and decay are part of nature's process and are allowed to proceed undisturbed. Our prescriptions of management are more rigid than in a national park, and this is one of the main reasons why we prefer to call them tiger reserves.

The springs have revived. The flowers bloom, no longer browsed by domestic stock. The ungulates have all the space they need and their population has increased. Tiger sightings are more frequent and there has been a baby boom. In Jaipur Zoo I failed to provide an environment for a tigress to rear her full litter of six cubs, and similarly with two tigresses, Radha and Vindhya, in Delhi Zoo, who gave birth to five cubs; but in Ranthambhor, in one of our reserves, a tigress has successfully brought up five cubs, thus creating a new record of success even in the wild.

You are sure to irritate a Bengali if you are undiplomatic enough to tell him that the Bengal Tiger is not the largest. Equally I will be unhappy if you say that Rajasthan is no place for the tiger to live. Nobody wants to admit that there are no tigers in his land, or that his tigers are smaller, less colourful or less ferocious than other people's. Now the tiger is claimed to be at his best everywhere. The concept of total environmental preservation has paid dividends, and the tiger has once again returned to these areas to the surprise of the new generation.

The enthusiasm of people from all over the world, especially the European children who saved their pocket money and sold tiger stamps, was so encouraging that we invited a few of them to come and see the results of their efforts. Particularly encouraging, too, was a note in *International Wildlife* (1975) from Hal H. Harrison and his father George who travelled extensively in the reserves under the sponsorship of the Wildlife Federation of America: 'In spite of the cryptic criticism of the non-official wildlife enthusiasts the boys on the firing line are doing a heroic job,' they wrote. I feel this compliment is a well-deserved reward.

The project has fully demonstrated that by tradition and training the Indian forester is much better equipped to administer a conservation project than anyone else; this view is shared by my friends the American foresters who manage wildlife habitats.

* * *

I am greatly encouraged by the response of the habitat, the tigers and their prey in the Tiger Reserves. It may be too early to predict the outcome of this effort, but it is surely not too much to hope that ultimately the tiger will be restored to a less precarious position than he is in at present. If I have made myself unpopular with some people in championing the tiger and not allowing them to exploit him for business ends, that is unimportant to me. I have the satisfaction of knowing that I have played a part in maintaining his dignity and not allowing him to be degraded to the status of a trophy, a guinea pig or a frog in a biology class. Some at least of the world's few remaining tigers are being neither hunted nor confined behind bars—they are free. As we become ever more aware of his value we may be sure that the tiger, his associate animals, and the tigerlands of India will continue to delight and inspire generations as yet unborn.

ARJAN SINGH

THE WAY AHEAD

Somehow the priorities of our conservation programme have become hopelessly muddled: the whole effort has got bogged down in political and parochial bickering. Wildlife is essentially an international subject. Migratory birds, for instance, pass through many countries, and India can do nothing to stop the slaughter of Siberian cranes in Iran, Afghanistan or Pakistan. A tigress was found with her head shattered by a bomb which someone had placed in her kill. No inquiry was possible because she had come from Nepal and died in the Indian border area.

Yet—the world being governed by humans—the administration of wildlife funds and the management of wildlife programmes inevitably devolve on national organizations and become subject to national restraints. The developed countries generously gave a great deal of money to save the tiger, but in India we claim Project Tiger as an Indian venture. Not only that: we have given over most of its administration to the various State governments, all of which are under strong political and financial pressures. Although Project Tiger still falls within the central sector, the Government of India has unilaterally committed the States to producing half the funds needed, and this they greatly resent, claiming that their budgets are not large enough even for ever-increasing human needs, so they cannot afford the luxury of preserving animals.

In a democracy wild animals cannot exist without the will of the people. Unless the people living on the periphery of a wildlife park derive some benefit from it, the park cannot exist successfully for long. At the moment locals do derive benefit, in the form of firewood and grass for thatching, but they also suffer considerable aggravation from wild animals, and even though compensation is now payable on domestic stock taken by tigers, it is often not available unless a subvention is paid to the official who assesses the claim.

One of the greatest difficulties is that the Forest Department is a commercial organization, charged with producing revenue. Yet by historical accident and administrative convenience it also has charge of wildlife. In the old days the two sides of the business ran easily in harness: the British administrators regarded the animals of the forest as a valuable source of recreation, and culled and conserved them accordingly. Today, however, the demands of forestry and wildlife pull in opposite directions, for by the Indian Wildlife Act of 1972 commercial exploitation is banned from any area designated as a wildlife reserve. Thus it is in the interest of the Forest Department to keep wildlife parks as small as possible, and officials naturally resist any attempt at expansion. The result is that most tiger reserves are no more than islands of conservation under siege by the human invader, and too small to carry a tiger population which will be genetically viable in the long run.

It is now clear that commercial forestry and wildlife management are mutually antagonistic. Every tree felled for money or for forestry management means the destruction of habitat for birds, insects, reptiles and small mammals. A clean forest floor, on which humans scavenge the remnants of fallen trees, is anathema to the wildlife manager, for the presence of humans in the forest inevitably disturbs the animals and leads to fires which devastate vast areas. The planting of quick-growing, exotic species such as eucalyptus is equally inimical to wildlife, for it produces no under-storey of shrubs and grasses on which ungulates can feed, or in which predators can find shelter.

For these reasons, commercial forestry and wildlife management should obviously be run by separate administrations, each with its own expertise and training schemes. In the past the scientifically-trained forester looked on wildlife as a kind of bonus, in the form of the sport provided by shooting; now, when he is called on to act as a wildlife manager, he simply does not have the requisite knowledge and training. He becomes an 'expert' simply through holding a certain rank, and for his knowledge he is forced to rely largely on outdated books written by hunters whose view of animals was hopelessly distorted by the fact that they were always trying to kill them, and that the animals' behaviour was altered by continuous pursuit. At the same time, to be involved with wildlife has become a status-symbol, and attracts an unhealthy amount of kudos.

What India needs is an equivalent of the United States's Fish and Wildlife Service—a specialist Government body, with its own scientific expertise. Instead, we have left the care of our wildlife in the hands of a commercial organization whose political masters are constantly demanding increases in revenue, as well as privileges for the voters on whom they depend to keep them in office. It is sometimes said that, but for the Forest Department, there would be no wildlife left; in fact it is a wonder that there is anything left *in spite of* the Department.

A specialist Wildlife Department would be able to work out what the minimum population of tigers in any one pool should be. The IUCN, as I have said, laid down a figure of 300; yet the fact is that most Indian reserves, as they now stand, do not have room for even half that number of tigers. A pool of 300 tigers would demand a minimum habitat area of between 2,000 and 3,000 square miles, and reserves that size are available nowhere in India—although the Sunderbans and Manas would qualify, if suitable extra areas could be added. (For comparison, the Tsavo National Park in Kenya covers 8,000 square miles, the Corbett National Park 225, Dudhwa 190.)

What is certain is that in any finite area there are limits beyond which the tiger population cannot increase. The great predators are essentially territorial animals, and although they show a good deal of tolerance if prey is plentiful, they cannot stand becoming overcrowded. If this happens, they automatically adjust their numbers, either by the superfluous animals moving out into other ranges (if this is possible), or by the tigresses curtailing their own breeding-rate.

In the twelve years since the inception of Project Tiger, we have seen a welcome increase in numbers—although it is hard to tell how

K. Ullas Karanth

much of the gain is real and how much theoretical. Bureaucracy requires that the number of tigers in any one reserve rises steadily from year to year, thus demonstrating the success and skill of the local administration. If the return for one year shows forty tigers, the next will probably show forty-five, the next fifty-one, and so on. Real numbers, however, may be very different. Thus the present official estimate for the Corbett National Park is 103, but people outside the bureaucracy say that the reserve contains no more than forty. In Simlipal the official total is sixty-one, the unofficial estimate—given by another forester—between six and eight.

Whatever the true gain, it may turn out to be no more than temporary, for the small pools of tigers which we have managed to save will inevitably run the risk of deterioration from inbreeding, unless we can substantially increase the space available to them, or link existing areas by forest corridors, so that the tigers can effect genetic exchanges on their own. The very least we must do is build up a body of expertise at darting and translocation, so that if all else fails tigers can be efficiently sedated and transferred from one reserve to another.

We do, however, need to think bigger than this. We need to accept the principle of complete environmental protection, with the tiger—our national animal—as a symbol of the apex of the biotic pyramid. We need to make wildlife reserves inviolate and free from all commercial intrusion by man. Above all, we need to think more in international terms—not only for fund-raising, but for the creation of bigger reserves. One International Biosphere Reserve, for example, could be created by uniting the present forests of Bahraich, Kheri and Pilibhit with those of Suklaphanta and Bardia in Nepal, taking in all the intermediate areas of human population. The problems would be formidable, I know; but I repeat that they must be tackled with imagination.

Above all, we must act with *emotion*. Wildlife can only be saved now by a crusade, and like all crusades this one must have an emotional appeal. For too long the tiger has been looked after (if that is the word) by the bureaucrat with one eye on the politician, and the politician with one eye on the main chance. The rigid scientific attitude which frowns on an intermixing of sub-species is no longer adequate; rather, we must respond with open hearts and enthusiasm.

I believe that we have an inherited duty to preserve all animals. If something has come down to us in nature, through evolution or the Almighty—whatever your faith may be—I am convinced that we should accept it and look after it. Just because a creature is of little material use to us, and does us no immediately obvious practical good, we have no right to let its kind die out.

If we consider ourselves as trustees of the planet, we must mend our ways and cherish everything we have left. Our insensate plunder of the environment has brought disaster on every front. Destruction of the forests has led to soil erosion in the hills and flooding in rivers lower down. Extensive quarrying has caused hillsides to collapse. Even the climate seems to have changed for the worse under man's baleful influence. Through persecution and negligence, we all but exterminated the tiger.

Animals should be allowed to live for the sake of evolution, not because of the use they can be to the master race: we must cease to think of them as beasts, and accord them the dignity which we consider essential in dealing with our own kind. Their physical functions, after all, are the same as ours. Not only should we conserve the armadillo, the only animal known to get leprosy; we should protect the badger (even though he is suspected of carrying tuberculosis to domestic stock) and all the other beings, great and small, with whom we share the Earth. Above all we should fight to save the tiger, magnificent king of the predators, and a wonder of creation.

VALMIK THAPAR

A TIGER'S KINGDOM

Ranthambhore is a tiger's kingdom. Even within its four hundred square kilometres there is great natural diversity in its topography and terrain. The two hill systems within the park create a series of valleys, most of which are occupied by resident males. Valleys like Kachida, Bakaula, Semli, Berda, Nalghati and Lahpur are also connected to each other through a network of ravines and gorges that provide day shelters for the tiger. Each of these valleys has its own nuances, which must be understood if one is to absorb the varying behaviour of the predators and prey that live in them. The lake of Gilai Sagar is a large marshy tract of water at one edge of the forest, a sharp contrast to the general topography of the park. Looming above it is another deserted fortress, Khandar. This stretch of flat open land is ideal for the neelgai—the great Indian antelope—found here in large numbers. It also attracts a few thousand bar-headed geese every year; with its low-lying hills it has evolved in its own special way.

The system of three lakes that stretch around the fortress of Ranthambhore is in sharp contrast to Gilai Sagar. Not only does it appear distinctive to the naked eye, the behaviour of animals, be they predator or prey, is very different from that found in the rest of the park. The animals have had to adapt to a life built around large tracts of water. Within the waters of the lake live several species of aquatic plants, fish, soft-shell turtles and over a hundred marsh crocodiles. Countless species of birds live and winter around these waters: greylag geese (migratory birds which fly into Ranthambhore from colder climes like Siberia for the winter between November and March), black storks, saras cranes, painted storks, ibises, stilts, dabchicks, pelicans, teals, grey and purple herons, egrets, cormorants, darters, osprey, fishing eagles, serpent eagles, to mention a few. This busy and thriving ecological system exists within an area of about five square kilometres. High grass surrounds most of the lakes and large herds of sambar and chital are found concentrated there—they are more dispersed in the rest of the park. Because the area is open and less dense, small groups of neelgai and chinkara, the Indian gazelle, frequent the edges. Sounders of wild boar are a common sight. For the leopard the area provides little cover and I have never come across this predator around the lakes. The Indian sloth bear ambles about, especially when the 'ber' fruit is in season. Over the last eight years the resident tigers have adapted their hunting techniques to this area, making it a predators' paradise today.

The sambar deer have probably adapted the best to life around the lakes. Normally regarded as a shy and elusive resident of the thicker forest, here they are found in their hundreds, congregating around the lakes but more often entering the water, wading and swimming, eating the succulent plants that grow within. Their life is water-based and even the colour and condition of their coats is in sharp contrast with those that live elsewhere in the forest. Sambar are not found in such large herds anywhere else. From November through to March is their courtship season, and stags crash their antlers into branches and grass as they prepare to court and fight for supremacy of the harems. Surprisingly, much of this activity takes place in and around the lakes. Their antlers fall off during April, May and June, and when the monsoon breaks they are in velvet, a living tissue that feeds the antlers till they are fully developed. They spend most of the wet season on the plateaus of the park. Their numbers become more concentrated around the water points as the heat intensifies and as water sources begin to dry up. Living on the aquatic plants in the lake is not without its share of danger. Adult crocodiles attempt continually to kill the deer, but are only successful when a young fawn strays into deep water. Adult sambar usually manage to escape by a great deal of kicking when there is any hint that the crocodiles may attack. But over the last few years four or five large crocodiles have played havoc with the deer, and many more sambars have fallen victim to them. This successful predation has not deterred the deer from their addiction to the water and its plants. Young fawns that grow up in the area follow their mothers into the water and somehow develop as 'water deer'.

Until the end of 1983 the sambar's only concern was with the few crocodiles that might lunge at them. In the water they were safe from tigers. This also changed with the appearance of a new resident male—Genghis—at the end of 1983. Sambar form the largest prey base for the tiger around the lakes and Genghis adapted his hunting technique masterfully by studying the sambar's addiction to the water. From the cover of high grass around the lake he would shoot out into the water, his speed and accuracy remarkable, causing total panic amongst the deer in the water. The sambar were unable to rush away easily, as they might do on dry land, and invariably young deer would fall victim to Genghis. His success rate was one in five, higher than tigers that kill on dry land. He remained around the lakes for a year, a ferocious predator. He would even take kills from the crocodiles when they were successful. But even during this period of ruthless predation, the sambar never left the waters of the lake. Genghis disappeared late in 1984 and we have never seen him again. But he left behind a hunting technique that the resident tigress, Noon, had picked up, probably from watching him in action. She now became the 'water killer'. The sambar remained undeterred. I wondered what might happen when her cubs were grown. They would naturally pick up this innovative technique.

Unlike the sambar, chital never enter the water. They graze in herds at the edges. The chital court and mate in Ranthambhore throughout the year. This activity reaches a peak during March and May. Occasionally they are chased to the edges of the lake by a tiger which has been sheltering in the high grass. But being exceedingly fast on open ground, they avoid predation more successfully than the sambar. This is true only of the lake areas. In the Nalghati and Semli valleys, an equal number of chital and sambar are preyed upon, because the sambar are much more dispersed.

Other prey on the lakes are wild boar, neelgai, the occasional langur monkey and peafowl. Large populations of peafowl live around the lakes and are attacked frequently, especially if young cubs are around. Cubs learn their early hunting techniques by chasing peafowl. Nocturnal animals like the porcupine and ratel also fall victim to the tigers.

Observation of wildlife around the lakes is a naturalist's dream. The area is picturesque, scattered with ruins and a lake palace, and the ever-imposing fortress of Ranthambhore. The terrain is open, affording excellent visibility. The first lake, Padam Talao, or 'the lake of the lotus', is adjoined a hundred metres away by the second lake, Rajbagh, 'the garden of kings'. Jogi Mahal, the forest rest house, sits on the edge of Padam Talao with its terrace serving as an excellent observation post. My diary record of 15th March 1982 states:

It is 8.30 a. m. and I have just returned from a morning drive. I sit out on the balcony of Jogi Mahal to have a cup of coffee. I watch a crocodile as it glides through the water. A pied kingfisher hovers above before dropping into the water like a stone, coming up again, seconds later, with a fish. The lotus is about to bloom, and most of the lake is covered with a spread of green leaves. The morning drive has been uneventful and I wonder what the evening holds. My gaze drifts to a herd of fourteen chital grazing on the lush green grass at the edge of the lake. The coffee arrives and I take my first sip watching this serene lake and its surroundings when quite unexpectedly a cacophony of chital alarm calls diverts my attention to a tiger that has charged the herd from the tall grass, startling them into confusion and successfully catching one. The suddenness of the attack has caught me by surprise. The next moment the tiger grips the neck of the chital and carries it off into the high grass around the lake. It looks like a doe. The whole event has lasted two minutes and the lake appears to be back to its former calm and serenity.

On innumerable occasions in 1984, from the terrace of Jogi Mahal, I saw Genghis and Noon charging through the water and killing sambar. Rajbagh, the second lake, is the sambar's favourite. You can spend hours observing deer, birds, crocodiles and tigers when they are around. This lake is half a kilometre away from Malik Talao, the third and smallest lake, and the only one to dry up in early summer. Small pools of water lie between them. The area behind Rajbagh is cool and shady, like a nursery for the sambar and chital. Mothers with their young take refuge there. Vehicle tracks swing all around the lakes and the area has the disadvantage of being the most frequented by tourists. Endless jeeps whizz about and if a tiger is present it can attract up to ten vehicles. But I knew this area when few tourists came and today if there are no jeeps it is still one of the finest places to be in. I have seen a tiger kill about fifty times, and on forty occasions it was in the lake area.

I once watched six crocodiles attacking a young sambar in the waters of one of the lakes, twisting and turning their bodies in an effort to yank off the limbs of the deer and amidst much splashing, tearing the carcass apart. Crocodiles are important scavengers around the lake. In May 1985 a large sambar stag collapsed and died in a sun-baked clearing at the edge of Rajbagh. The dead stag provided a focal point for four days of feeding activity in which we were able to study at close quarters the natural hierarchies and interactions of the scavengers of Ranthambhore. Early on the first morning we found the area around the carcass criss-crossed by crocodile tracks. The reptiles, whose teeth are not very sharp, had only managed to tear a chunk of flesh from the sambar's neck before retreating into a nearby pool. Crocodiles find it difficult to eat on land. It was a lucky break for the vultures; white-backed, Egyptian, king, griffon and long-billed vultures all congregated to feed, tearing at the carcass through the opening made by the crocodiles and all the time flapping, kicking and fighting amongst themselves. The next morning a pair of jackals arrived on the scene, approaching cautiously and then charging the vultures, trying to find a few mouthfuls. Late in the afternoon four wild boars arrived. They approached aggressively, forcing the jackals to retreat. The boars annexed the carcass and gorged themselves. They established themselves firmly as the dominant scavengers. Vultures and jackals kept away and even the crow and tree pies found it difficult to snatch a few scraps. The wild boar is a tough animal. Even tigers keep their distance from large adult boars, especially if the tiger is at all inexperienced as a hunter. On the third morning evidence on the dusty soil indicated the arrival of a tigress. She had picked up the few remnants of bone and flesh and dragged them thirty metres away to some bushes where she had chewed and licked the bones clean. The area around the lakes had become the arena for some of my most revealing encounters with tigers.

Noon has been a resident tigress of these lakes since early 1983. She must be now between seven and eight years old and weigh close to 180 kg. Noon has certain distinctive features. After many hours of observation you learn to recognize her facial markings over the eyes and cheek. She also has a broken right lower canine, easily visible when she yawns. Her character is different from the other tigresses we have observed. She hunts around the lakes at any time of day and probably bases her technique on the movement of the deer at the lake's edge. Unlike Laxmi, she is aggressive around a carcass, sometimes mock charging the jeep, forcing us to keep a greater distance. She tends to favour long bursts of speed in pursuit of prey, again adapting her technique to the open land around the lake. Laxmi, on the other hand, prefers to hide in thick cover, stalk her prey slowly and attack over much shorter distances.

The Semli valley is about ten kilometres from the lakes. Its own special characteristics made it possible for us to document the early life of the cubs, as a network of ravines, lush and green, were all accessible by jeep. We could not have driven so close to the cubs in the lake area. Laxmi, the resident tigress of this valley, is nearly twelve years old and this was the second litter of hers that we had seen. She is slightly heavier and a few centimetres longer than Noon. She is also much calmer and less aggressive. Her territory has little open ground and she has adapted her hunting techniques accordingly. Unlike Noon she is much less active between midday and five o'clock.

Tigers vary enormously in their individual natures, and their behaviour is governed by their immediate environment. The Nalghati tigress is about the same age as Noon but more heavily built. Her hunting area is the Nalghati valley, a stretch of land seven kilometres long, surrounded by steep cliffs and hills with thick lush gorges. Her periods of intense activity are the early mornings and late evenings.

At one end of her area in a deep gorge lives the only human inhabitant of the park, Kawaldar Baba. He spends most of his time in meditation in a small cave on a cliff face. He has in a way merged with nature. On innumerable occasions he and Nalghati have encountered each other at close quarters, and there has been no threat or sign of fear on either side.

The two resident males in our study differ considerably in size. The Bakaula male is bigger and heavier than Kublai, who is sleeker and more athletic looking. We are unsure of their ages but think that they are somewhere between seven and nine years old. Kublai is more diurnal than the Bakaula male, who tends to disappear into cover for most of the day. Both have thick ruffs of hair around their cheeks.

There are an equal number of sambar and chital in Ranthambhore. The total figure is about 9000. There are about 2000 neelgai and 700 chinkara. About 1,500 wild boar roam the forests. Thousands of langur monkeys and hundreds of thousands of peafowl complete the prey base in the forest for about forty tigers and probably an equal number of leopards. Langurs are the only primates living in Ranthambhore. They spend much of their time in trees, safe from a tiger's attack, and cackle-bark in alarm at the sight of a tiger or a leopard. Following their eyes is a great help in pinpointing the exact location of a tiger. Few tigers, but many more leopards prey on stray domestic livestock on the fringes of the forest. Over the years most of the leopards have been pushed to the fringes by increased tiger activity within the park. The first leopard that we had seen in fifteen years eating an adult sambar ate furtively, having killed it in Laxmi's home range, wary of the tiger's presence. A sambar is three times the size of the leopard, so the leopard must have been forced off his feet to grapple with his prey. Young tiger cubs might be vulnerable to a leopard's attack, but usually leopards keep their distance. The strong scent of the tiger families must act as a deterrent. The same is true for the hyena. In Ranthambhore hyenas are found singly or in pairs but never as a pack. The total population of hyenas in the park must be about twenty-five, and they are exceedingly difficult to see. Again, their more regular activity is around the fringes of the park, scavenging the carcasses of livestock that may have died naturally.

Traditional hunting tribes had always killed jackals and could sell their skins for a hundred rupees. Because of this the jackal population was devastated, and although a total ban on hunting jackals was imposed five years ago, it is still rarely seen. In fact we see more tigers than jackals. Once between Padam Talao and Rajbagh I encountered two jackals. They had just killed a tiny chital fawn and were in the process of tearing it apart. In minutes scores of vultures circled the area, some settling on the trees nearby. Dozens of crows watched the jackals cawing noisily. Attracted by the sounds, Noon emerged from a bank of grass two hundred metres away. She watched the vultures and crows for a while, then strode forward to investigate. As she approached, the jackals abandoned the fawn and trotted in circles around her. Quickly she picked up the carcass and walked off to a cover of grass to eat, leaving the jackals in a frenzy of frustration. Recently, Noon killed and ate a passing jackal. This was our first observation of such an incident.

There are about 100 to 150 sloth bears in and around the park. The tigers and sloth bears keep their distance from each other. Unlike leopards, which are submissive with tigers, and sometimes even fall victim to them, the bear remains unperturbed and I have witnessed bears walking past tigers without any great concern. The tiger remains alert, watching them move away, but does not adopt an aggressive posture. Only once did I see Genghis, who was a particularly aggressive tiger, charge an unsuspecting sloth bear; both bear and tiger slapped each other with their forelegs and finally the bear retreated. But both seemed to respect each other's presence.

There are about 150 marsh crocodiles in Ranthambhore. Their

main diet is fish, supplemented sometimes by deer. Around the park are two river systems, the Chambal and the Banas. Both have healthy populations of crocodile and gavial, and the surrounding area is protected as a national park in order to keep nesting colonies safe.

It is difficult to get accurate figures of the larger mammals and reptiles that live in and around the fringes of the park. At the height of the summer, usually in the first week of June, an annual census is conducted, primarily for the tiger but also encompassing other animals. The forest is divided into about a hundred compartments and each compartment is manned by two forest personnel. For several days all observations around the water holes of that particular compartment are recorded on a map of the area that has to be filled in. So a tiger's movements, its pug marks, are all filled in on the map, and each day's count of other animals is also recorded. It is like filling in a big jigsaw puzzle for approximately seven days. The information finally reaches the forest headquarters and an 'animal concentration' map of the park is put together for each of the seven days. The accuracy of such a method depends on the reliability of the forest personnel. It can therefore only be successful when every forest tracker and guard is highly motivated, very knowledgeable and experienced in the census area that he is assigned. He must be fully trained in the recognition and analysis of the pug marks of all the forest animals and must be familiar with the day-today activities and movements of the tigers within his area of forest. But this is too much to expect. Recruitment into the forest service is not based on a man's interest in the tiger, or in wildlife. Many forest guards join the service because they have to earn a livelihood. Census operations tend therefore to be treated casually and suffer from large margins of error. The figures I have quoted are mere approximations of the total picture.

Late in August 1986 I visited Ranthambhore for a few days. The transformation was extraordinary. From a dull yellow at the height of the summer the forest had become a lush green. The grass was high, providing dense cover. The weather was humid, the insect life buzzing. An endless croaking of frogs and the gurgling sound of water pervaded the atmosphere. It was a great time for the reptiles. Cobras, kraits and vipers were at their most active, taking their toll of the frogs. The Indian monitor lizard streaked around the nests of birds and preyed on snakes. Many migratory birds flew in to feast on the endless dragonflies that filled the skies. Crocodile eggs were getting ready to hatch. All vehicle tracks were like streams and any depression in the ground had turned into a water hole. The network of ravines and what had been dry stream beds gushed with water. The deer and antelope were invisible, most of them on the higher reaches of the park. There were few signs of tigers. The forest guards revealed that they had occasional glimpses of Noon around Jogi Mahal, which gave us hope that her cubs had survived the wet season. The rains had not been as good as they should have been, the water levels not as high as normal. We would not know how serious the drought was until later in the year. For now, Fateh and I could only speculate about October when we would rediscover all three families. I was convinced that Noon and her two cubs would provide us with the most rewarding encounters around the lakes, with their superb visibility and thriving prey. In fact, documenting the lives of three tigers around the lakeside, in these startling surroundings, would be a total delight. I knew that since my favourite area is the lakes, I would concentrate on Noon and her cubs from October on. Fateh would have to spend more time with the Semli and Nalghati families. In any case the lakes would be the first area to become accessible after the rains: the valleys of Semli and Nalghati would not open up for vehicles until weeks after we had been able to approach the lakes.

Early one morning, the day I was to leave, a huge storm came crashing through the park: thunder, lightning and torrents of rain. Visibility was down to little more than a metre. Within half an hour, as suddenly as it had come, the storm disappeared, leaving in its wake a blue sky and the early morning sun. What a transformation! The quality of light was special, the fragrance in the air magical. The leaves dripped with water and the forest glistened as if it had been cleansed. Torrents of water cascaded down the hillside and waterfalls sprouted from steep cliff faces and the edges of the fortress. These would be the last heavy showers before the rain wore off. It was time to leave for the station. As we turned a bend in the road that runs to Sawai Madhopur a tiger cut across our path. As it paused briefly to look at us, we realized that it was Noon. She ambled off towards the ravine where her cubs were born, disappearing into a profusion of green.

VLADIMIR TROININ

THE YEAR OF THE TIGER

> The hunter risks danger . . . but man is far more dangerous to the tiger.
>
> —Vsevolod Sysoev

Planning one's life is a stab in the dark. Three decades ago I was the first among graduates of the Irkutsk Agricultural Institute's game and wildlife department to arrive in Vladivostok, firmly determined to work, not in the Taiga, but at sea—to observe and understand the lives of our biggest relatives on this planet, the whales. I was convinced that the ocean would be my second home for a very long time. But fate ordered otherwise: before embarking on my first voyage I had to work for several months in the Taiga. In the October days of the now distant year 1960, I traveled on the Taiga's Amba River to the far-off, snow-covered expanses of southern Primor'e. Planning a recreational hunting facility—that was my job description.

I wandered long in solitude through extraordinary lands. I read tracks. I evaluated woods. I considered where huts could be built for people who would hunt in this Taiga many years after me. And on the very first morning of the first day, the first animal track that crossed my path was that of an Amur tiger. I did not know at the time that this was a sign. After all, *amba* means 'tiger'. And I had no idea that at dusk of that very day I would observe a young, striped predator battling a wild boar.

I didn't understand fate's message. Upon completing the job, I went to sea. I worked for a few years in the Arctic, the Antarctic, in the tropics. I observed whales, and I wrote a book about them. But fate was insistent, and returned me once more to the Taiga. At first it sent me in search of herbs and health. Then it granted me yet another personal encounter with the striped wanderer. Fate lured me with the secret life of the forests of Primor'e, and on every trip into the Taiga I came across the trail of the Amur tiger.

And a quarter of a century after that day on the Amba, fate delivered a surprise. In 1986, the Year of the Tiger, six of those striped creatures settled for quite a long time on a peninsula next to the city in which I live, and their tracks in the snow revealed so much about their lives that my observations, as well as former encounters, make it possible for me to tell this story.

It was a snowy winter in the south of Primor'e. On the Muraviov-Amurskii Peninsula, so much snow fell that even on the southern slopes of hills, where in a normal winter one could walk in regular shoes, the snow was well above the knees, and in places the wind blew drifts up to one's waist.

In a snowy winter, any animal has a hard time finding food. To husband their strength, the inhabitants of the forest follow trails through the deep snow. A multitude of paws and hooves stamp out narrow, winding roads in convenient places: along sparsely wooded slopes where the view is good and the walking easier, along the courses of rivers and streams, where there is more prey for predators and other food for nonhunting beasts. A collective path through a winter wood is a convenient road. Even tigers prefer to take the beaten route in snowy winters. But when the need arises, these powerful animals clear their own trail; and such a road, followed several times, is evidence that the beast has taken a long-term liking to a place. It was onto such a path, heading off towards the town, that I walked on a March day that year.

It was not the first time I had followed a tiger's trail near the city. Solitary striped wanderers had occasionally come to the peninsula before, usually on winter days. They would settle in the region of Bogataia Griva or under the ridge of Skalisty Khrebet, catch stray dogs for a week or two, and, without disturbing humans, return again to the expanses of the Taiga. But this time something told me there was more than one hunter here. I had a feeling that the tigers had settled in for the duration of winter.

All day I followed tiger tracks. I found a spot where the beast had rested several times and a path by which it had descended from the pass to the valley of Bolshaia Sedanka. I followed its main trail, along which it had walked about four times to houses on the outskirts of town in search of dogs. This was a very large tiger—judging by its tracks, it weighed well over two hundred kilograms. By the end of the day I also knew that the beast was not alone. A somewhat smaller tiger had occupied the territory near the botanical garden, in the valley of Malaia Sedanka.

On the next day and for three days after, my wife, Olga, and I walked from dawn to dusk through deep snow. By the end of our labours we had discovered that five tigers had descended on the Muraviov-Amurskii Peninsula and had carved it into five distinct hunting zones. The two largest tigers had settled in the spacious valley of the Bogataia River. A younger one was roaming closer to the outskirts of town. Two others, in whose tracks I had walked on the first day, had taken spots right by the city itself. We did not know at the time that an expectant tigress had settled on the peninsula as well.

The six had not come to the area in vain. I am certain that some 'scout' had told them that a multitude of stray dogs, gathered into bands large and small, had been hanging about near the city. An abundance of such easy prey would allow the tigers to live quite well through a difficult winter, and in the first month—as evidenced by their tracks in the snow—they had calmly and efficiently cleared the woods around town of stray dogs. But then came a day when the inevitable occurred: a city dweller, walking his pug dog on the outskirts of town, was frightened nearly to death when an enormous tiger stole his pet out from under his nose. Horrible rumours began to steal through the town.

The tigers continued economically—and without troubling their human neighbours—to catch their dinners. At twilight they would rise from their lairs, go to outlying homes, drag pigs and domestic goats from sheds, and kill dogs. They called on summer homes and roamed at night near resorts, where one can always find leftover food. They missed no delicacy—they even checked out chicken coops. One tiger made several trips to the town dump.

The rumours of the tigers' invasion continued to grow. It was said that they had killed an old woman, eaten a young sailor, and become so brazen that they regularly looked into house windows. Those rumours were all untrue.

Much was written at that time in the newspapers about the Vladivostok tigers. Animal specialists held meetings to discuss what needed to be done to protect the city dwellers from danger. Special brigades of experienced hunters were organized. Huge amounts of money were spent to survey the peninsula by helicopter. In some places, specialists set traps with live dogs in cages as bait. By the end of the Year of the Tiger, five striped predators who had never touched a human being and who had cleared the peninsula of stray dogs were killed.

People killed those huge, beautiful, intelligent cats who had reminded us, yet again, what poor masters we are. Only a bad manager allows hundreds and thousands of domestic dogs to grow wild, wander scattered through the woods, and become outlaws, destroying the fowl near cities and towns. Only lazy slobs throw tons of food in the dump. People offer predators easy prey, leaving domestic fowl, calves, and pigs in unguarded, flimsy pens. Tigers, who lived for centuries with thrift, long ago adapted to human mismanagement.

One of the six who came to the peninsula was a very young tiger, a youth in human terms. After the dogs abandoned the hills in fear, it walked, after dark, to the field houses of the agronomy institute, next to the reservoir. One night, it crept up to a yappy little dog, who had barked till nearly daylight for no reason at all, and killed the dog to prevent it from interfering with its work. Then it stood a moment in thought, and began to examine the rows of wooden storehouses. It found the most convenient shed, one with a turnbolt. With one movement of its paws, it turned the bolt, pulled the door open, and went in. In the far corner a goat cowered in terror . . . The next night, it tore the rotten boards from the chicken coop. The night after that, it crossed the road and inspected several nearby yards. During the day, its hunger satisfied, the tiger rested far from human dwellings, near the source of a remote spring, on a small rock under a cedar tree. It was an ideal resting place: almost all day long the sun warmed the tiger's side, the beast had a wide view, and right nearby there was a crossing by which a hasty getaway could be made if the need arose. The young beast had considered everything thoroughly. But one quiet night it strayed from its usual route—it heard, far off, a dog yelping in a cage.

Another of the tigers was very large. An old hand, it made its lair over a sharp bend in Bolshaia Sedanka, in a sunny spot protected from view by a thick wall of undergrowth. For several days it rested only thirty metres from the old forest road where city dwellers frequently went for walks both during the week and on Sundays. Right next to the road ran a beaten ski track. How many times did the tiger cautiously raise its head? Perhaps, when it heard voices, it merely twitched its cars. Perhaps it did not react at all to nearby sounds. This beast knew human habits well! Who would go poking into an impassable hole, and what for? Who would even think to notice deep impressions next to the path and ski track, where a tiger had several times jumped across human roads on its way to the outlying houses of the city? This was the same tiger who had indulged in the mischief of snapping up a dog from under its master's nose. This was the tiger whose trail I had followed in the first days of March.

Reading tracks in the snow, I was amazed and impressed by the animal's reasoning and behaviour. For several days it had found its food within the city limits. Only twice did it leave its beaten track. Once, to watch and wait for a stray bitch, which it killed without touching her puppies as they slept in a hollow log; and again, to climb to a high, open place, where it stood for a long time, then sat, admiring the extraordinary view of the river valley and reservoir.

And it sat in one other place for a long time, motionless. Why? I sat down next to this place. I sat as it had sat, looking in the same direction it had looked. It was a wonderful place in the deep channel of a spring at the bottom of a far ravine. I felt with my back and shoulders how the peace was maintained by nearly vertical slopes, bristling with oak woods. The silence was captivating and one had no desire to stir. My view was directed at a narrow stream bed, covered over with a fantastical weaving of gnarled trunks and thick grapevines. A place for reflection. That tiger was killed from a helicopter.

Towards spring another tiger, the largest, began to make its home not far from the military storehouses. Watchdogs disappeared nearly every night. A special brigade of experienced shooters stood guard for many days. But before it was killed, and despite a prohibition, inexperienced hunters twice shot and wounded it. For thirty-five days, without disturbing a soul, it carried a bullet in its flesh. Then it was shot and killed.

Amur tigers! At one time these powerful predators were called Siberian tigers. To them belonged the expanses not only of the Amur River basin: occasionally the first explorers also found tigers on the southern shores of Lake Baikal. In those years, in the Tunkinsky Valley at the upper reaches of the Irkut River, tigers sometimes 'tended' the wild boars. They also owned the Selengi Valley and visited the upper reaches of the Olekma. The farther the Russians moved towards the Pacific Ocean, the more often they encountered the powerful and freedom-loving masters of the Taiga lands. These beautiful and strong beasts demanded respect: they were the living symbols of those lands. The Taiga without tigers is an empty Taiga. Not without reason does the emblem of Irkutsk, Russia's largest province, depict the *bábr*, a beast that today is mysterious to us, carrying a sable in its teeth. The *bábr* can be interpreted to represent the tiger, which, like the sable, symbolizes the wealth and beauty of the land.

The striped wanderers are the largest and most long-legged of cats, unafraid of deep snow and hard frost. Long ago they came to the Lena, disturbing the residents of Yakutsk. An old inhabitant of the town of Kumor, which is a hundred kilometres north of Baikal, told me that his grandfather (this was in the beginning of the century) unexpectedly came across a tiger in the dead of winter. A Taiga native, the grandfather was not afraid. He examined the extraordinary animal and concluded, 'He wasn't hungry at all—he had everything in order!' The last 'frost-resistant' tiger was killed in 1905 on the Aldan Plateau.

From Aldan to the Sea of Okhotsk, the tigers' way was barred by the high peaks of Dzhugdzhur, but the wandering creatures made it to the shores of Okhotsk from the south. Overcoming the icy emptiness on winter days they travelled to the Shantary Islands. Even at the beginning of this century, they frequently investigated Sakhalin. At the southernmost end of their enormous range, they reigned over what today are Primor'e, Korea, and all of northeast China. So it once was . . .

It took humans just over a century to edge out the tigers, leaving them the eastern part of the Amur River basin. People seem to have unwittingly interfered with the natural development of a new subspecies of the world's most powerful hunter, the Siberian tiger. But our Amur tigers aren't so little either. They now live on the wide stretches of Primor'e, at the very south of the Khabarovsk Region, and in China, on the left tributaries of the Ussuri River. Occasionally they travel to the northern end of the Korean peninsula. No one knows how many Siberian tigers there once were. Today there remain only slightly more than three hundred Amur tigers.

Our Amur tiger is distinguished from the other six subspecies by its great size: an old male can weigh in at three hundred kilograms, though this is rare. It has the brightest colouring and the thickest, longest fur. Like its cousins, it has a powerful and flexible body, a beautiful, large head, and strong paws with tough, springy cushions that allow it to stalk its prey silently. With its terrible claws, the tiger instantly inflicts fatal wounds.

On that October evening so long ago when I saw, close up, a young tiger battling with a very large wild boar, I noticed that it mostly used its front paws: it would fall on the boar from the back or side, and give it such a blow that the heavy pig would topple over on its side. Several times the boar managed to pin the tiger to the ground and inflict severe wounds with its teeth, but even in that position the tiger tore the boar's thick skin with all four claws and its teeth at once. It was clearly the weaker of the two, but it had the advantage of more and better weapons. What's more, it could strike from nearly any position.

Fearing that the beasts would notice me while in their fighting mood, I left the battlefield—making no abrupt movements—and didn't witness the outcome. On the morning of the next day, I went back and thoroughly examined the spot. The snow was packed down in three places, and everywhere were clumps of fur and spots of blood. Several thin oak saplings and birches were broken. The killed boar was horrible to look at! Its belly was torn, its neck and throat one big wound; its left front paw was yanked out of its joint, its face covered with welts, and its nose cut off as if with a razor. After the boar had been skinned and I had the chance to examine it, I was most struck by the scratches on its hard skull bone—each more than a centimetre deep! Yet its killer was not a large tiger but a three-year-old that weighed just over a hundred kilograms and that took some serious damage itself.

For half a day I 'unravelled' the tracks of that hunter. While descending from the cliffs, it had heard boars romping in the undergrowth. About twenty metres from the herd, the tiger lay down in the snow. It then began to creep forward and, after about fifty metres, lay down again to wait for the boars to come within springing distance. Judging by the boar tracks, there were piglets in the herd, but apparently this was a spirited young tiger: it chose the prey most beyond its strength, and proved its superiority through struggle.

I followed its fresh trail—sprinkled with little drops of blood—several kilometres away from the place of battle. It had gone off to rest, occasionally falling in the snow, licking its wounds. It had learned a good lesson the hard way, but its ferocity and its confidence were strengthened.

My first tiger was inexperienced and overly adventurous. Normally, these able hunters fell their victims with the first blow. Nature was generous when it created them. Strength and deftness, elegance and endurance, intelligence, *sang-froid*, patience, and thrift are harmoniously combined in the tiger. Tigers have phenomenal strength. I have heard of a tiger jumping one and a half metres into the air with a fifty-kilogram animal in its teeth! And I have seen a birch tree, fifteen centimetres in diameter, broken down by a tiger that had pulled a Manchurian deer out from where it was stuck between two trunks. Its elastic muscles produce nearly unbelievable strength in a moment of need. I've seen a dog—a fat one, about thirty kilograms—fly five metres from the blow of a tiger's paw. Try to throw a fifteen-kilogram rock that distance with both your hands. I remember the genuine surprise of a young sailor describing a Vladivostok tiger. One night the sailor had shot at it, sitting in the beam of his headlights only fifteen metres away. 'It was sitting there like a cat, on its hind legs, its front legs straight. It had a dead sheepdog at its feet. It looked at me. I shot! Without swaying, the tiger leaped up in one second to a height of two metres or more, turned its back to me in the air, and . . . disappeared into the dark!' This was the tiger that was later wounded and, in another month, killed. It weighed two hundred fifty kilograms.

Tigers have wonderful control of their bodies, not only on firm ground but in air and water as well. There is a well-known incident in which a tiger overtook a young bear in a stream and killed it as it swam. In 1960, on the Amba River, tracks told the tale of a tiger that killed on the fly a wild goat that had jumped high into the air. A swift jump allows felines to approach their prey in an instant and quickly escape danger. Here are examples I have observed: it took a rather fat-looking cat a little over two seconds to get to its prey, three metres up a tree trunk and one metre out on a limb; it took a leopard, running away from me, no more than four seconds to ascend a steep nine-metre cliff.

Nor is endurance a weak point for the tiger. One of the tigers that came to the Muraviov-Amurskii Peninsula in 1986 ascended and descended six steep mountains in very deep snow on its first journey to the city, never stopping to rest. About ten years ago, I exhausted myself following a fresh tiger trail more than forty kilometres. I rested several times; but the animal walked the whole distance at an even pace without a single stop. Even nowadays, with hundreds of hunters stealing the tigers' main prey—boars, wild goats, and deer—I don't believe that tigers have much of a problem finding food. In search of food, a hungry tiger can travel a hundred kilometres or more, and on a journey that long something is bound to show up. Tigers never rely on chance. I am certain that those intelligent creatures are able to calculate precisely, and far in advance, where and when they will be able to find prey in their chosen territory.

The tigers know the Taiga well: they know their victims' habits and can orient themselves so they quickly locate spots from which they can watch prey at a distance. Coming to the peninsula, they never missed the treeless heights of Skalisty Khrebet and Bogataia Griva. They always stood there for a long time, studying the land and calculating the routes of the stray dogs. I believe they took a good look at the daily routine of the townsfolk. Why else would a tiger lie on one of those heights and look for so long towards the city?

This ability to calculate in advance told half a dozen tigers at once to come to Vladivostok in 1986. They knew for certain that they could live out the winter and eat their fill there. What is more, the tigress correctly judged that she would be able to feed herself and her little ones. She gave birth to two cubs in an ideal place: closer to Artem, in a dense ravine, no more than an hour's journey to a fence behind which were many spotted deer. In need she could go in the other direction, to garden plots where, as summer approached, there were always many dogs.

One of the Vladivostok tigers displayed exceptional powers of calculation, *sang-froid*, and knowledge of human habits. It lived quite comfortably for several days in the midst of human activity. I found its lair in a neglected gully, only a hundred and fifty metres from the rostral column marking the entrance to the city. As it rested there during the day, hundreds of cars raced by, newlyweds had their pictures taken by the column, people enjoyed themselves in the restaurant across the road. The tiger knew that these people were occupied with their own affairs and were unlikely to go poking around in a damp and unpleasant ravine. It slept on the only dry spot. It knew also that the cowardly city dogs, sensing the presence of a terrible beast, would run from the place without a sound, and that in the event of danger, it could quickly cross the road and vanish into a vast and thick forest. On the other side of its lair and right next to it were many wooden homes. Here, every night, it killed stray dogs. A couple of times it pulled bothersome sheepdogs from their collars. Once it allowed itself the amusement of yanking a couple of boards from the wall of a shed and tasting suckling pig. It then frightened a little dog out of its wits as the latter cowered behind a fence, trying to bury itself in the ground. The next day, the poor dog shook the whole time and refused to leave its pit.

The incident with the dog surprised me. Tigers don't like dogs, but they prefer fresh dog meat to any other food. It's not a matter of flavour; it's a matter of utterly incompatible natures. Nervous, noisy dogs irritate the calm, silent cats with their barking and yelping. They stick their noses into the tigers' hunting grounds, causing anxiety and panic among the tigers' prey. When a tiger destroys a dog, it not only feeds itself, it also brings order to its property. In some tigers the dislike of dogs becomes a passion. Several years ago a Terneiski tiger, dropping in regularly on one settlement, killed twenty-five adult dogs in a month but touched no puppies. Perhaps some parental instinct was at work.

Late January to February is a time of love for the Amur tigers. Hitherto having taken no notice of—indeed, running from—one another, tiger and tigress now spend more than a month together. They have a wonderful, romantic time. Wherever the loving couples live, one can see evidence of playful leaps and of much lying and rolling. In spring and early summer, the cubs are born.

Tigresses are diligent mothers. Their cubs are always fed, safely sheltered, cared for, and kept out of danger. A mother tiger makes short work of any beast that poses a threat to her children. She might give even their papa a whack if he happens to set foot into the nursing area, which is now off limits to him.

As they grow, the cubs get to know the Taiga and its laws under affectionate but strict tutelage. Intelligence and rational behaviour are always encouraged; mischief immediately earns a good slap. As they mature, the cubs begin to understand that Mama is the best teacher. My old friend Viktor Beschastnyi, an experienced student of the Taiga, made a detailed study of the teamwork of a pair of tigers. A tigress and her grown cub together followed a small herd of boars above the valley of the Maximovka River, then separated. The cub climbed the ridge and walked alongside the herd at a distance of nearly three kilometres. The tigress followed the boars from behind. The herd crossed a saddle and turned sharply to the right in the next valley. The tigress and the cub, who at that moment were one-and-a-half kilometres apart, turned after the herd simultaneously. What signalled the young hunter to turn at just the right moment and in the right direction? He couldn't see or hear the boars. The tracking ended with a successful catch by the cub. Even before his mother's arrival, he came out in front of the herd and killed a piglet. Unravelling the tracks, Viktor came to the conclusion that the tigress knew the area well, had considered all the possible variations in the boars' path, and had somehow related this information to her son—who then followed her instructions precisely. She followed behind only as a safeguard. I believe she was satisfied with her child's capable work.

When the cubs are almost fully grown and can stand up for themselves, the tigress, after feeding them well, leaves them for a few days to prepare them for independence. In February 1979, together with the biologist-ranger Viacheslav Bazyl'nikov and his laboratory assistant, I tracked tigers on the southeast slopes of the Sikhote-Alin Mountains. I regret to this day that I was elsewhere, following another tiger, when Bazyl'nikov and his assistant came across the trail of a tigress and her four cubs. The tracks in the snow told them of the cubs' behaviour as they experienced what was possibly their first taste of independence.

The tigress had left them for a time in a hazel grove, descending the slope to a place where there was much prey, and had killed a large boar. After her breakfast, she had gone to the river and drunk, and that night brought her family to the prey. There was plenty to eat, and she left her cubs for two or three days. Bazyl'nikov and his assistant, Vladimir Ivanov, closely examined the spots where the cubs had stayed near the food. In the snow were traces of playful battle. The little ones had played at hunting each other. Making a large circle, one of them had crept up on its brothers and attacked them. One cub was especially clever. It had decided to test its strength on a rather large boar and had begun to pursue it. The tracks of the cub and the boar were fresh. Bazyl'nikov hurried uphill—they were somewhere nearby. The bright and bold tiger cub apparently sensed that it was being followed: it turned and waited in ambush. Bazyl'nikov saw it, about twenty steps away. He later wrote in his journal: 'The cub was a bit larger than a big sheepdog. It was concealed in shrubbery at the edge of the clearing, waiting for me to come into the open. We looked each other over well. The expression on its face was at once a bit cowardly, bloodthirsty, and curious.' The cub didn't intend to attack the man: overcoming both terror and excitement, it wanted to observe. It left first, when Bazyl'nikov took barely a step in its direction, and returned to its brothers and sisters. It must have told them about the encounter, for the family then retreated from any possible danger.

On almost the same day, a large tiger observed me. Towards evening I crossed its fresh trail, went to my winter hut, and, as I opened the door, felt on my back an intent and unfriendly gaze. I turned carefully, and through ten-power binoculars I thoroughly examined the sparsely wooded side of the hill, illuminated by the evening sun. The tiger lay a hundred metres or more away, pressed like a cat against a gnarled birch tree, its big head resting on its paws. The binoculars allowed me to examine its wide, whitish sideburns; its ears, slightly pressed back; the bristle of its long whiskers; its intent gaze. I remember that a chill went up my spine.

That's the way we are: our interactions with tigers are a problem for us. When we come across one of them, our terror tells us that an empty-handed human is helpless to stand against the powerful predator. But why must we necessarily stand *against* it? Why does terror incline us to battle? Why are we certain that malice guides our neighbour of the Taiga when a human crosses its path? Facts speak to the contrary. For many years, thousands of hunters, geologists, foresters, tourists, mushroom gatherers, berry pickers, and vagabonds have been passing through the tigers' land, all returning home alive and well. Tigers tolerate the presence of humans in their home, and only occasionally remind us with a menacing growl that we have crossed the boundary and are in their territory.

The tiger is unquestionably a dangerous and powerful hunter, but equally unquestionable are its nobility and restraint regarding humans. You won't find ten genuine accounts of tigers attacking humans in all the years since World War II. But who can say how many hundreds of tigers have fallen victim to poachers in that time? Now, animal specialists have become proficient at trapping them, and nearly every animal that they catch alive, they kill. The all-powerful press is cultivating the belief that relations between humans and Amur tigers—who apparently interfere with our lives, work, and leisure—are possible only with the aid of clever traps and firearms. These people deal with the predators without ever going near them! And we have somehow forgotten that, in our motherland, something used to happen that is unknown in the history of any other country: bold Russian men took on tigers with their bare hands! They took on, it's true, cubs most often; but our Amur tiger cub is in fact the equal in strength and weight of the Bengal tiger, which even today is approached only on the back of an elephant. Hunters today have been spoiled by technology, but in the past, there were true tiger catchers, indeed: Averian Cherepanov, Miron Popov, Luka Gorbunov, Vasilii Glushak, Rodion Kozin, the renowned Bogachev brothers, the Kulagins, the Trofimovs. In the summers from 1930 to 1940 alone, Russian tiger catchers in the Iman basin (today the Bolshaia Ussurka) took on with their hands forty tiger cubs, each weighing about eighty kilograms, and nine full-grown tigers. And no one today even remembers how many tigers Vasilii Glushak alone took on with his bare hands. In the fifties, Vsevolod Sysoev wrote of these fearless men: 'No, the *bogatyrs* [mythic Russian heroes] have not died out in Russia!'

From childhood we are instilled with the idea that any predator is always dangerous. This is true. But how should one act upon coming in close contact with such a terrible beast as the tiger if one is armed? Even in my fellow rangers, encounters with tigers have immediately provoked a reflexive reaction: in an instant the carbine freezes to the shoulder. My colleagues, however, managed to retain their humanity in those situations. Here is what my classmate Sergei Kulikov told me in a letter: '[T]he devil told me to look around. Next to me, on the very bank of the stream, sits a striped cat, looking at me sweetly! I don't know what the tiger thought, but I only became able to think once I was looking at it through the sight of my carbine. How long we contemplated each other in this manner I don't know, but the tiger, continuing to look at me, lay down most calmly, stretched out its front paws and raised its head high, though its ears moved constantly, and it seemed to me that each moved independently—to the right, to the left, backward, forward. Probably at that moment the biologist in me took the upper hand: I wanted to photograph the creature. I even remember thinking: There won't be another such opportunity. But I confess, at first I hadn't the courage to lower my carbine and pick up the camera. I had a Zenith-E at the time, with an advancing lever. Holding the carbine to my shoulder, I opened the camera case with my left hand, took off the lens cap, cocked the camera, and pressed the button. (The picture did not come out well; it was out of focus.) As soon as the tiger heard the click, it pricked up its ears, its expression changed, and to this moment I don't know how I kept myself from shooting. But the tiger turned away and began to listen intently to the silence of the Taiga, looking at the path along which I had approached the stream, and along which it had silently followed me. It showed me its back, listened to the Taiga, and then in one leap, from a lying position, instantly disappeared into the bushes.'

In his encounter with a furious tigress, the biologist-ranger Anatoly Kuzmin showed courage. During a boar hunt, a rather large cat began sneaking up on Kuzmin. Thinking that a leopard was playing a game with him, he decided to give the animal a scare, and approached the knoll where it was hiding. Right in front of the knoll he was paralysed by a terrible roar: it was as though an enormous tiger had risen from the ground and fallen upon him! Raising his carbine to his shoulder, he realized that it was a female. And it was her cub that had been stealing up on him. Three times the tigress came within springing distance of him! He found within himself the strength to explain to the enraged mother that he would not touch her cub, and asked her pardon for invading her Taiga. Her third warning was especially terrible: she was clearly preparing to spring. Bur Kuzmin did not shoot, and they parted ways. He later said to me: 'There's a beast! She could easily have attacked me from behind. But she came to me in the open. She warned me! She protected her offspring without cowardly devices, and she let me know whose Taiga I was in.'

Biologist and Taiga expert Viktor Korkishko had an incredible experience, too. For an hour he conversed, unarmed, with a tiger! A big old beast approached him unexpectedly as he was writing in his field journal. After a warning roar, it closed the distance between them to five metres and stopped. As it approached, Korkishko waved his hands, anxiously crying, 'Get out of here! Go away!' Then, seeing the animal's determination, he made the only correct decision and began speaking in a confident tone. Later he wrote in his journal: '[A] completely meaningless text: "Kitty, you're so smart, big, beautiful, strong, what do you need me for? Let's go our ways. I'll go that way, you go the other, I won't do anything to you, I'm unarmed. Go away, please, let me go."' The tiger stood there quietly. But as soon as Korkishko stirred and began to withdraw, the tiger came closer. At a distance of two metres it began to show its disdain: it yawned, looked about, rubbed its cheeks on the bushes, bit off twigs and chewed them. Again, Korkishko made the only right decision; he began to try to buy off the tiger with objects. He gave it his hat, goggles, rope, vest, a pack of cigarettes. He backed away gradually as he proferred his gifts. I believe this was the only possible tactic, and here's why: more than once, I have had occasion to observe the behaviour of a female cat when teaching the law of the land to a weaker rival forced to live in the same territory. Every day the strong cat, choosing the right moment, would force the weaker one into a corner, where it would have to fall on its back and howl, humiliating itself. Conscious of its complete power over its opponent, the first cat would display its contempt. Only a few centimetres away from the degraded cat's claws, she would stand sideways to it and begin to gaze at something in the distance, yawn, and wash herself. The humiliating show could last for quite a while. And if, God forbid, the cornered cat should forget itself and howl with malice, it would immediately be slapped in the face. It is necessary to humble oneself, albeit with dignity, in such a situation. And Korkishko was consciously and gladly driven into the corner by a powerful, crafty beast. Yawning and looking aside, the tiger was vigilantly observing his behaviour, and it received the gifts with pleasure. But suddenly, it closed the distance between them to one metre and began to watch him fiercely and decisively! He found the courage within himself to stay calm—he didn't take his eyes off the tiger, even when its nose was within half a metre of his legs. The game of nerves continued until dark. The tiger appreciated Korkishko's collected behaviour, and allowed him to cross to the other side of the river. 'The last picture: it stands looking at me. Soon I heard the sound of a snow tractor.'

It happened that the driver of the snow tractor was Dmitrii Pikunov, one of the most experienced of rangers and someone who had studied the Amur tigers for many years. He didn't believe Korkishko's story. Only the next morning, after studying the tracks, was Pikunov convinced, as he himself put it, of the 'incomprehensible probability' of the event.

On a sunny December day in the Year of the Tiger, a helicopter hung over the Muraviov-Amurskii Peninsula. The pilot made tight arcs over hills and dove deep into valleys. Sometimes the helicopter hovered as the people inside it strove not to miss a single sign attesting to the presence of a tiger. The last of the striped cats that had come to the peninsula was being hunted. For nearly a year this experienced creature had escaped retribution; perhaps it counted on living here comfortably for another winter. But the tiger didn't know and couldn't appreciate how determined people can be when they get carried away.

On that day, Olga and I, wringing sweat from our hats, were walking to a place, known only to us, where a week ago that tiger had rested. We walked and were happy to see no trace of stray dogs. The tigers had cleared the peninsula of those four-legged bandits. And the forest had clearly come to life this winter: there were more wild goat tracks than there had been a year ago. But all the same, our hearts were heavy. The fate of the last tiger had been decided.

The helicopter rushed over us, hovered, began to descend, and landed in the treeless space on the pass. We saw someone run out of it and hide in the bushes. Then two more people came out, armed, and the firing began. We knew for certain that the tiger was not there, and could not be. What were they shooting at?

We went up to the helicopter. It turned out that the pilot was shooting just to pass the time—at stumps. The person who had run into the bushes had motion sickness and was nauseated. The third was a forester whom we knew somewhat, the owner of a conspicuous pair of shoes, the tracks of which we had noticed several times. He greeted us with surprise: why would we be panting our way up this steep hill? I said that they were shooting rifles very close to a road. 'So what?' was the reply. The road was quiet here. They were shooting in the other direction. And the tiger would be killed in any case.

The helicopter flew off. We descended into a steep ravine and soon found the tiger's tracks. The path was well smoothed—the animal had walked it more than once, had walked without stopping even where the snow was up to its belly. The tiger's road is not easy! And we lightened our labour by walking in its tracks and thus transformed the elegant path into a carelessly torn-up fissure.

The path took us up to the heights. Here the tiger had walked, choosing places where the trees were sparser and the view broad. Olga said, 'Look how beautifully and smoothly it walked, and how we're disfiguring its trail. A shame! We're destroying the harmony. Let's walk next to its tracks.'

It was beautiful here! Gold-and-grey trunks grew from the white snow. The transparent blue of the sky was tangled in the bare branches. Green cedar and fir needles in the depth of the ravine skirted the foot of the hill in a wide ribbon. And far in the distance continued the trail of the Amur tiger. I recalled the words of the old Udegei: *The Taiga has a great soul. The tiger is part of that soul. . . . If you kill a tiger, the Taiga is weakened. If you kill all the tigers, the Taiga will lose its soul entirely.*

PART 3

TIGER SCIENCE AND CONSERVATION

GEORGE B. SCHALLER

MY YEAR WITH THE TIGERS

My wife Kay and I were sitting on the veranda of our bungalow looking out over the forests and meadows to the hills of central India shimmering in the heat of the early afternoon. Several axis deer grazed a hundred feet away and peafowl rustled among the leaves of the forest floor. Suddenly a doe and her small fawn ran, barking in alarm, from the undergrowth. Thirty seconds later a tigress appeared in their tracks. She looked at the doe, then at the fawn which had strayed off to one side. All three froze, oblivious of our presence. At last the fawn broke and attempted to rejoin its mother by dashing past the tigress. In a fraction of a second the tigress had uncoiled her 300 pounds in flashing, bounding pursuit. She pounced on the fawn, muffling its dying bleats, and carried it into the brush. Casually, the doe began to forage.

I had come to India under the auspices of Johns Hopkins University to study big game, in particular the tiger and its relation to the deer, antelope, and other prey species. Until then, the tiger had been studied mostly along the sights of a rifle. The lore that emerged from these necessarily brief encounters attributed to the tiger fearsome characteristics of ferocity, speed, cunning and invincibility. Sporting teams and military units unhesitatingly choose the tiger as the symbol of their prowess. And so too, lately, has the American advertising industry seized on the tiger to sell everything from automobiles to hair dressing. But this lore, while entertaining, provided very little detailed information on the animal's habits.

My hope was to learn about the tiger's true nature, to become acquainted with the animal as it lived undisturbed in its jungle realm.

To do this, I roamed the forests alone and unarmed, observing tigers on their hunts, and watching them on moonlight nights from the cover of tree trunks and blinds as they feasted on their kills.

The area I had picked, after much searching for a suitable study area, was Kanha National Park in the state of Madhya Pradesh. It is a small park—not much more than 100 square miles—but quite remote and unspoiled. In the forests of this sanctuary it is still possible to recapture India's past, when wildlife was abundant almost everywhere, when domestic livestock had not overgrazed the range, and when tigers were not the rare creatures they have become elsewhere in the country as the result of decades of ruthless slaughter. Here, too, the tiger lives under optimum conditions, with its three necessities of life—food, water and cover—all in good supply.

The tiger, an adaptable animal, is found in many of the forested parts of India, Burma, Thailand, Laos, Malaya, Sumatra and in Siberia, as well as in a variety of habitats from hot, humid mangrove swamps, reed beds and rain forest, to dry semi-desert shrub and cold coniferous forests where the snow may lie several feet deep during the winter. Naturally it varies somewhat in its behaviour from area to area. However, I feel that my observations in Kanha Park are fairly typical of those tigers in India which have not been constantly persecuted by man.

The method I had chosen for studying tigers was not always easy. Tigers are of a shy and retiring nature, and they prefer to go about their daily routines inconspicuously. They travel largely by themselves and tend to stay in cover. They avoid man by stealing away or by hiding in the grass. Sometimes when a tiger resented my intrusion, it growled a warning or gave a coughing roar to indicate its anger. However, there was usually peace between us, even when I once inadvertently approached a sleeping tiger on a rock and we locked eyes at a distance of just three feet.

But the compensations in my method outweighed such disadvantages. I found that tigers are easy to recognize individually at close range by the distinctive patterns of black markings above each eye. I soon learned to distinguish the individual tigers in my study area, and some of these in turn probably knew me on sight. In particular, one tigress with four cubs provided me with endless hours of pleasure, for as the weeks and months passed they grew quite used to my presence.

The quest for food fills much of the tiger's life. It usually hunts alone, padding steadily along at two to three miles per hour, over forest paths, up ravines and through tall grass, eyes and ears alert, trying to sense some unsuspecting victim. Much of the hunting is done between dusk and dawn, when wild pigs, deer and other hoofed animals are out feeding and can be stalked most readily. But when hungry, or when a female has cubs to feed, a tiger may hunt at any time of the day.

The tiger's seemingly unbeatable array of weapons—its acute senses, great speed (but over short distance only), strength and size, and formidable claws and teeth—has given many naturalists the impression that the tiger can kill at will, that it lives in a sort of animal Eden with nothing to do but pluck, so to speak, the fruits of the forest. My experience shows quite the contrary—a tiger has to work quite hard for its meals. During the hot season at Kanha Park, when some 800 head of big game concentrate in less than five square miles in order to be near the few remaining waterholes, I observed tigers who sometimes hunted for several nights in a row without being able to kill a single animal. Conditions for the tiger must be just so. It must have some cover to be able to creep close enough to its victim before trying to surprise it, during the final rush of about thirty to eighty feet. The prey species also have their own defences, particularly an acute sense of smell and a speed greater than that of a tiger. I estimate that, for every wild prey killed, the tiger makes twenty to thirty unsuccessful attempts.

Take this example of a frustrating morning in the life of a tigress: I first became aware of her when a jackal yipped and raced along, closely followed by the tiger. She chased this scavenger for a full quarter of a mile, but was easily outdistanced. She then retraced her steps, spotted three swamp deer along the forest edge, and stalked them. However, the deer scented her and barked shrilly, facing the direction of their hidden enemy. Knowing herself discovered, the tigress rose and walked away, moaning softly as tigers often do after an unsuccessful stalk. Seeing seven swamp deer in the distance, the tigress anticipated their movements and hid at the edge of some high grass. When only forty feet separated the hunter from the hunted, the lead doe gave a shrill bark and the herd wheeled and scattered. The tigress rushed out, swiping the air with her forepaws in a futile attempt to reach one of the animals. She then strode off, while the deer trotted contemptuously some seventy feet behind her, knowing that the tiger, deprived of the factor of surprise, could do them no harm.

By carefully checking the age of nearly 200 remains of deer killed by tigers, I found that many were either young or old. Proportionally few were in the prime of life—and natural selection tends to save those deer with the sharpest senses and swiftest feet. Exceptions are females that are about to fawn, as their bulging abdomen robs them of agility.

In the case of the gaur, a huge type of wild cattle in which the bulls may reach a weight of over 1500 pounds, tigers prey mostly on the calves, taking as many as half of the young born in any one year. But once the gaur are grown, tigers tackle them only infrequently, showing great respect for the sweep of their horns. The tiger cannot afford to make the slightest mistake in attacking a potentially dangerous animal. There are records of gaur and buffalo goring tigers and of wild boars disembowelling them with their tusks.

With so many factors arrayed against them, it is not surprising that tigers may go hungry for several days. Sometimes they are reduced to eating langur monkeys, frogs, bird eggs, crabs, even berries. Once I watched a tigress pounce into a thicket in an attempt to catch jungle fowl, five of which flushed cackling around her. Any meat, including carrion no matter how decomposed, will do. And they are not above robbing a leopard of its prey. In one instance, a leopard killed an axis deer fawn near me, but that night a tigress with one small cub appropriated the carcass.

Taking India as a whole, domestic cattle and domestic buffalo are the most important items in the tiger's diet. With the wildlife in most forests decimated or eliminated, the tiger has taken partly or wholly to living on livestock, which is easy to kill and readily available. At least 10 per cent of the cattle grazing in Kanha Park and on its fringes are killed by tigers each year.

Once a tiger has made a kill, its life revolves around the carcass until all meat, viscera and soft bones have been eaten. The tiger sleeps nearby, perhaps under a shady clump of bamboo or partially submerged in a forest pool. At intervals throughout the day it returns to the remains, sometimes for a snack and at other times to charge with a roar at any imprudent crows or vultures that may have descended to the kill. The tiger often hides the carcass in a ravine or paws grass and dirt over it to prevent the scavengers from stripping the meat off the bones. The number of days spent beside a kill depends, of course, on the size of the victim. An axis deer weighing 125 pounds may last two to three days, a 300-pound swamp deer five or six days, before the last rotting scrap has been devoured.

I estimate that the average adult tiger needs about 15 to 20 pounds of meat per day, or roughly 3½ tons per year, to remain in prime condition. However, as only about 60 to 70 per cent of each prey animal is edible, the rest being bones and stomach contents, each tiger has to kill about 4½ tons of animal every year. Thus, the annual needs of a tiger would be satisfied by some thirty of India's scrubby cattle or by about seventy adult axis deer. Actually, tigers have to kill more than this in most areas, because they are frequently chased from the remains by man, especially if they have attacked a domestic animal. Where tigers are persistently shot at over their kills, they have learned never to return for a second meal.

When all the meat has been devoured, the tiger again begins its nightly rounds, probably covering about 15 to 20 miles of terrain between dusk and dawn. Most adults hunt within a definite range using certain routes and resting places more frequently than others. One tigress I came to know at Kanha roamed over about twenty-five square miles of forest, but most of her hunting was done in about eight square miles. However, the ranges of some tigers, especially in areas where game is scarce, may be considerably larger.

A tigress will readily share all or part of her jungle beat with other females and with a male. However, an adult male probably will not tolerate for long the presence of another male within the boundaries of his range. At least at Kanha, I noted that only one male was resident in the whole central part of the park, and another was once a casual visitor, whereas I knew of three resident and at least six transient tigresses. Such transients appeared to be tigresses in heat who had abandoned their range in some distant forest to wander widely in search of a mate. They passed through my study area, particularly from December to February, stayed perhaps a day or two, then disappeared. A male probably mates with any female in heat, resident or transient, found within the boundaries of its territory. The territory thus functions as a spacing mechanism, decreasing competition between males for the females.

Perhaps no aspect of the tiger's behaviour has been as misunderstood as its social life. They are often said to be totally solitary, except during mating. It is true that tigers generally hunt alone, but it is also quite common for them to meet briefly during the night. Tigers possess quite efficient means of long-distance communication. The main one is a call signifying, 'Here I am'—a roaring *aa-uuu, aa-uuu*, repeated again and again with emphasis on the first vowel. In the stillness of the night this sound carries for well over a mile. Both male and female tigers raise their tails at intervals when walking along and spray a mixture of scent and urine on trees and bushes, leaving a powerful odour which in some instances I could discern for as long as three months. The odour undoubtedly helps tigers to track and find each other in the forest, in addition to serving as calling cards.

Tigers also share kills. Once, for example, a tigress loudly called her distant cubs to dinner by sending her *aa-uuu* rolling over the forest. From afar came an answer. More than an hour later another tigress joined the family in their feast. On another occasion a tigress with four cubs and a tigress with one cub remained together for at least two days at a kill.

The male tiger who presided over the section of Kanha Park that I was studying was a huge, amiable fellow with a scraggly ruff surrounding his face. He was a gentleman in the true sense of the word when it came to dealing with the females and cubs that shared his range. The tigresses in turn appeared to trust and like him, and they made no effort to keep him away from their cubs, even fairly small ones. Once a total of seven tigers—two tigresses, four cubs, and the male—shared a bullock. On another occasion the male visited the female with four cubs at their kill. Although he was obviously hungry and the meat must have been tempting, he lay down twenty feet away and waited patiently until everyone had finished before taking his first bite. Intermittently the cubs rubbed their faces against his and sinuously moved their bodies along his head and neck—the typical friendly greeting between tigers. He remained with the family until morning, then resumed his rounds—solitary but certainly not unsociable.

The most lasting social bond is between a tigress and her cubs. After a gestation period of about 105 days the cubs—there may be only one or as many as seven, but the average is two to three—are born blind and helpless under some rocky ledge or fallen tree. There the mother suckles them and perhaps brings them meat. By the age of six weeks they leave their shelter to accompany the tigress to the kill for the first time. From that day on they never again have a permanent home, but live like nomads, moving from covert to covert and from kill to kill. Their life is spent waiting quietly for hours and days in some hidden spot for the tigress to return to lead them to a meal.

I first encountered the tiger family I was to become particularly well acquainted with when the four cubs were about four months old and the size of setter dogs. The mother had killed a bull gaur in a shallow ravine and I watched the family throughout the night from behind a tree at a distance of 100 feet. During the hours of darkness the cubs mostly ate the carcass and slept, but when the sun replaced the moon they began to play. They raced in a single file, stalked each other through the grass and wrestled vigorously. They explored their surroundings, sniffing at leaves, looking into holes, occasionally pouncing on something—perhaps an insect. As the cool of the morning gave way to the searing heat of the day, the tigress arose, emitted a series of low grunts—evidently meaning 'Follow me'—and led her brood to a shady bower nearby. For five days the family camped by the dead gaur until it was gone. Then they disappeared. Eleven days later I again saw the tigress, walking across the meadow at mid-afternoon dragging a freshly killed gaur calf by the neck. The calf's mother paced nervously back and forth at the edge of the forest but lacked the courage to attack the tigress. Stopping now and then to adjust her grip, the tigress hauled her kill out of sight into a grassy ravine. There she apparently ate a little and then rested. An hour and a half later, she reappeared and padded steadily off into the forest. When darkness fell, several swamp deer barked hysterically as the tigress returned, bringing her cubs to the feast. Then came the sounds of tigers at a kill: the crunch of bones, the grating of teeth cutting skin, the occasional growl or cough of a minor dispute between the cubs. The cubs had their own pecking order, with the largest taking the choicest feeding spot and the smallest sometimes having to wait for its meal until a place was vacated. (When food is scarce, the weaklings probably die. It appears uncommon for a tigress to raise as many as four cubs to adulthood. But mine seemed to be making it.) By next morning the family had totally finished the gaur calf, and, led by the tigress, they filed away to a copse of trees near a waterhole about half a mile distant. I had weighed the calf earlier, when the mother first went to fetch her cubs, and now I checked the few remaining scraps of bone. The tigers had eaten a total of 85 pounds during the night.

While the cubs are small, the tigress sometimes takes a respite from her hunting duties and rests even though the larder is empty. The family I watched once ate an axis deer during the night, but on the following evening, instead of going out to hunt again, the female just lay in the grass on her back, legs up in the air, while the cubs wrestled with each other all around her. One of the cubs climbed over her stomach, another followed and she cuffed it lightly, then a third draped itself over her face and gnawed on an ear. All the next day the animals lolled indolently in the deep grass, but at dusk the tigress set off alone on a hunt. Only a little over an hour later she was able to kill a buffalo that had strayed from the village—a lucky find which provided the family with meat for another two days.

As the cubs grew older, nights of leisure were a thing of the past for the tigress. She had great difficulty in securing enough food to satisfy her ravenous brood. When the cubs had reached the age of one year—and were as large as a St. Bernard—they required the equivalent of one axis deer per day and it was almost too much for the tigress to handle. Repeatedly she invaded the cattle enclosure in the nearest village at night until, in desperation, the villagers removed all their livestock from the area. She also stole a lamb we were fattening for our Christmas dinner by clawing her way into our shed. And unable to obtain enough food at home, the male cub—by then considerably larger than his female siblings—set off on his own. He was inexperienced but determined. One day he entered the village and secured a pig; and on another night he demolished our chickenhouse and ate all five occupants. However, he remained in contact with the rest of the family and usually showed up when there was something to eat.

Until the age of one year the cubs seemed to have no actual experience in hunting big game. But now one or more sometimes accompanied their mother on her nightly prowl. She in turn appeared to make a definite effort to provide them with experience in the art of killing. On one occasion she felled a buffalo without injuring it, then stepped back while three of her cubs attempted to kill it. They were so inept that the buffalo shook them off. Once again the tigress pulled the animal down for the cubs, and finally, biting almost at random, they managed to dispatch it. Obviously the cubs had much to learn before they could survive on their own. Even at the age of sixteen months the female cubs were not proficient at killing. However, the male cub, which weighed well over 200 pounds and by trial and error had gathered experience on his own, had become quite adept. Soon the bond between the cubs and their mother grew tenuous. Every two or so days she dropped by to see them, and undoubtedly provided an occasional kill—just enough to tide them over from partial independence to the complete independence they would reach by the age of two years. By that time she would probably have a new litter of cubs.

This then is the life of the tiger. To kill to live is its main concern, and much of its energy and time is devoted to the task of securing a meal. Like all cats the tiger presents a violent contrast between action and indolence, gentleness and courage, shyness and persistence, and like all cats it possesses a certain exalted indifference that makes any attempt by man to enter the world of the tiger both difficult and challenging.

Yet it is worth the effort to try. Partly—perhaps mostly—there is the beauty of the beast. To see a tiger striding on velvety paws across a meadow of yellow grass, with the fading rays of the setting sun harmonizing with its tawny, black-striped coat—self-assured, a picture of barely contained power yet lithe grace, the very symbol of physical beauty, strength and dignity—is surely one of the greatest aesthetic experiences in nature.

JOHN SEIDENSTICKER, R.K. LAHIRI, K.C. DAS
AND ANNE WRIGHT

PROBLEM TIGER IN THE SUNDARBANS

In August 1974 a young male tiger moved into a populated area in the Sundarbans, the delta of the Ganges, and killed one woman and a number of livestock. Rather than destroy the animal the Forest Directorate decided to capture it, using immobilizing drugs, and release it in the Sundarbans Tiger Reserve. This was successfully done, but less than a week later it was found dead from wounds evidently inflicted by another tiger. The authors discuss the implications of the incident, the publicity it attracted, and the changes in public attitudes.

In the early morning of August 2, 1974, a tiger was reported to have killed a woman near the village of Jharkhali, 70 km south-east of Calcutta, in a reclaimed part of the Sundarbans in the Ganges delta. The tiger was sighted repeatedly over the next few days, and people were very alarmed. On August 7, the State Wildlife Officer, R.K. Lahiri, arrived by motor launch to investigate and report to the West Bengal Forest Directorate, the authority responsible under the 1972 Wild Life Protection Act.

The tiger was roaming in a densely populated region of a large delta island, a mosaic of paddy fields, villages and a large central mangrove marsh used as a fishery. The area had been part of reserve forest lands until about 1955, when it was reassigned for the resettlement of refugees. The remaining reserve forests on the southern end of the island were about 3 km from the villages where the tiger was observed. A narrow belt of mangroves along the Matla River on the west side and the central fishery area provided the tiger with shelter and cover but no large mammalian prey. Further south, in the reserve forests and in the tiger reserve, axis deer (*Axis axis*) and wild pigs (*Sus scrofa*) occur in good numbers.

The body of the dead woman had not been eaten. Lahiri[9] reported that there was a goatshed less than a metre from where she had been sitting, and, as the incident occurred in the early hours of the morning, 'the circumstances suggest, that the woman may have fallen victim by accident'. The tiger had also killed dogs, cattle and a chicken, but in most cases had not been able to drag away or even feed on its kills, being driven away by the shouting and disturbance people created after each kill. The tiger did not defend its kills, nor did it return to the sites.

The Sundarbans forests are one of the very few remaining places where tigers still occasionally kill people for food,[13] and vivid accounts enrich many old shikar books. The reasons have been a continuing subject of speculation and there have been investigations recently in Bangladesh by Hendrichs[7] and in India by Chaudhuri and Chakrabarti.[5]

Occasionally, large and potentially dangerous mammals become isolated or stranded in unsuitable or populated areas from which they must be removed. In the past there has been little alternative but to destroy them, Now, however, effective chemical restraint procedures and equipment make it possible to capture the animal, transport it and release it in a suitable area. Although widely used in Africa and North America,[17] this management technique had never been used on free-roaming large mammals on the Indian subcontinent.

After this particular incident extreme pressure was brought to bear on the responsible officials to destroy the tiger immediately. But, as the evidence indicated that it was neither a man-eater nor an incorrigible man-killer, it was protected under the 1972 Wild Life Protection Act. The Forest Directorate decided to capture and translocate it by darting, using methods worked out during a tiger ecology study in the Nepal *terai* by Seidensticker *et al.*[15] Once immobilized the tiger could be examined and, if physically fit, moved to the Sundarbans Tiger Reserve. If not fit, it could be kept and treated in captivity under controlled conditions until cured, and then released back into the wild. Seidensticker was invited to assist and flew down from Nepal.

Capture and Release

K.C. Das and R.K. Lahiri were joint directors of logistics and field operations for the Forest Directorate mission. Two motor launches were used for transport and communications with Calcutta and Port Canning, five hours away.

The tiger's approximate location was soon established. The reclaimed populated area is maintained with a system of bunds (dikes) which provide people and livestock with a trail network to move through the mangrove marsh and paddy fields. An intense monsoon storm and unusually high tides had breached the bunds in many places, and the region was mostly inundated by the daily 18-20 ft. tides.[6] However, pugmarks made early in the morning of the seventeenth led to one of the dry thorn-covered hummocks in the central marsh, near the spot where a cow had reportedly been killed four days before, and where next day we found a cow's scapula. We decided to try to hold the tiger here away from the villages while the fishery owners repaired the dikes so that the water level could be lowered. Meanwhile we worked on developing the conditions needed to dart the tiger successfully.

Over the next nine days, we tied three bullocks on the hummock as baits, secured with nylon rope to prevent the tiger from dragging them away. He killed and ate the first two, and also two dogs that came to feed on the carcasses. A bullock that wandered into the area on the night of the 19th was attacked but escaped along the narrow bund.

After four days of observations from the tong, a boat with a secure hide for observation and darting was moved into position less than fifty feet from the baiting site. It was clear that the tiger would only come to the baiting site in the dark, making a night darting necessary, so we built a 15-foot-high machan 100 feet away for a spotlight.

The tiger then managed to pull the remains of the second bullock free and drag it into the mangrove scrub out of sight. So the third bait was tied. The water level had been lowered, conditions were right for darting, but the tiger returned to feed on the hidden remains of the second bullock. He did not come to the third bait until just after midnight on the twenty-sixth, when he rushed and killed it. We darted him ten minutes later and found him immobilized 60 metres away on a mud bank in the mangroves.

He was a young male not fully grown. We examined him thoroughly and found no physical defects, and, as in the previous days that we had been observing him he had appeared to be behaving as normally as could be expected, there was no apparent reason to keep him in captivity. He was therefore put into a zoo transfer cage and moved to the launch, and by daylight we were already on our way to the release site in the core area of the tiger reserve. After about two hours the effects of the drug wore off and he could push against the bars. We took great care to avoid any undue disturbance during the journey.

At the release site, the cage was carefully moved ashore, turned on its side, and the door pulled open with a rope from the launch, thus avoiding the need to immobilize the animal again. After a few minutes the tiger walked out into the deep mud and went off into the thick mangroves behind the cage.

The follow-up of the operation could only be done by periodically checking the area, as radio-location was not permitted in the tiger reserve. On the afternoon of the twenty-eighth the tiger was seen back in the transfer cage, but was gone by morning. On hearing this the Chief Conservator, Mr Roy Choudhury, asked Lahiri, Das and Seidensticker to return to the site, where we arrived at midday on September first. We found the tiger dead in the mangroves, twenty metres from the transfer cage. There were numerous tiger tracks around both the cage and the carcass, and the freshest tracks, which had been made in the few hours before the last high tide, and many of the older ones were larger than those of the dead tiger. The carcass was already badly decomposed. The right side and abdomen had numerous maggot-infested puncture wounds, the thoracic cavity had puncture wounds, and there were multiple puncture wounds in the hips and right shoulder. Only the extremities had been fed on by scavengers. A small monitor *Varanus salvator* ran away as we approached. No other large predators occur in this region, and we could only conclude that this tiger had died of injuries resulting from an encounter with another tiger.

Observations of tigers over the years have provided only the most preliminary picture of the social organisation and land-tenure system of this big cat. Schaller's work in Kanha National Park[13] greatly advanced our understanding, but the ability to predict requirements can only come from intensive studies using modern field techniques such as radio telemetry.

Scattered through the old *shikar* literature are reports of tigers fighting and even killing one another,[1,2,4] but without any critical data such as the exact conditions and the social role of the individuals involved. Fighting has occurred also in confined situations such as zoos and wild animal parks[12, 16] and is a factor that has to be considered in captive management. But Schaller's work in Kanha and McDougal and Seidensticker's observations in Nepal[10] showed very little overt aggressive interaction. In these free-ranging tiger populations, where individuals were presumably familiar with one another, fighting was unusual; avoidance and restraint were the rule.

The outcome of this translocation was unfortunate, but the Forest Directorate's monitoring effort turned it into a learning experience. Obviously we are a long way from understanding the complexities of injecting an 'outsider' into an existing social structure, and under present conditions this ignorance is critical.

As long as people and tigers live side by side, tigers will occasionally wander into or become trapped in areas from which they will have to be removed. One alternative is to capture and move them to a zoo, or some similar confined situation. This is perhaps a notch above killing the animal outright, but with the large number of tigers already in captivity, the recent successes in captive breeding, and all the evidence indicating the continued decline of tigers in the wild, we did not consider this a logical or even legitimate approach.

The intensive management required to ensure the tiger's survival in the wild will involve manipulating the environment to enhance living conditions as well as applying precise techniques to the tiger population directly. We can anticipate the problems of maintaining genetic variability in isolated populations of what was once a widely distributed species[3] and, from our knowledge of population processes in other solitary big cats[14] we can also anticipate difficulties in maintaining breeding populations in the small isolated sanctuaries which are the basis of the current tiger preservation programme.[8, 11]

Our lack of understanding of these fine-tuned behavioural factors and mechanisms hampers our efforts to preserve these great cats, and action must await the results of carefully planned long-term research. Meanwhile there will continue to be problem tigers, and recent events in Orissa and Assam, where problem tigers have been shot, point to the urgent need for guidelines if such tigers are to be preserved in the wild. Suitable release sites with good natural prey populations will have to be identified, often perhaps in forest reserves rather than in existing tiger sanctuaries. Managers and research workers must cooperate, and an infrastructure is needed for developing plans and for reviewing and incorporating new information, as it becomes available, into all field operations.

The publicity achieved by this Sundarbans tiger incident was considerable and the public attitudes interesting. The fact that this tiger killed a woman was immediately reported in the press, and the debate about the tiger's fate, the capture, the translocation, the tiger's death and the ensuing discussions were all fully reported. Our file from Calcutta and New Delhi English-language newspapers alone includes eighty-one articles. For over six weeks the operation was constantly before the public.

Unfortunately no detailed survey of public attitude was made before and after the operation, but it seemed clear from the reporting that shifts were occurring. At the beginning the pressure for the tiger's destruction was intense; by the end, the death of 'Sundar' seemed to be lamented, and, of course, there was much debate on how the operation should have been conducted.

The publicity was unplanned and we tried to avoid it; it added considerable pressure to a difficult undertaking. But for the larger conservation perspective it was important. At first the arguments raised were the old anti-predator ones, but at the end, the non-politically motivated accounts seemed to be moving towards a position in line with today's environmental approach, with a growing realization that the fate of this great cat truly lies in man's hands and in man's ability to provide for its ecological needs in a man-dominated environment. Ultimately, the tiger's survival in the wild depends on public awareness and attitudes towards the larger issue, that of man learning to come to grips with the constraints of the land and developing a realistic rapport.

Reference

1. BAKER, C. 1886. *Sport in Bengal: How, When, and Where to Seek it.* London.
2. BRANDER, A. 1923. *Wild Animals in Central India*, London.
3. BUECHNER, H. and MARSHALL, D. 1975. *Threatened Ungulates of North America.* In Proc. Symp. on Endangered and Threatened Species in North America. Wild Canid Survival and Research Center, Wolf Sanctuary, St. Louis (in press).
4. BURTON, R. *The Book of the Tiger*, London.
5. CHAUDHURI, A. and CHAKRABARTI, K. 1972. *Observations on Tigers: Wildlife Biology of the Sundarbans Forests.* Divisional Forest Office, 24-Parganas Division, Calcutta. 18 pp. (mimeo).
6. FOSBERG, F.R. 1971. *Mangroves and Tidal Waves. Biol. Cons.* 49: 38–39.
7. HENDRICHS, H. 1972. *Project 669 Tiger: Study of Man-eating Problems in the Sundarbans.* In World Wildlife Yearbook 1971-72, 109–115. Morges, Switzerland.
8. INDIAN BOARD FOR WILDLIFE (GOI). 1972. Project Tiger. New Delhi.
9. LAHIRI, R.K. 1974. Report on the Investigation of Depredation Caused by Tiger in Parbatipur Mouza, under Basanti Police Station. West Bengal Forest Directorate. 5 pp. (typed).

10. McDOUGAL, C. and SEIDENSTICKER, J. 1976. Predatory Behavior of Tigers (*Panthera tigris tigiris L.*): Ecological and Sociological Aspects. 120 pp. (typed ms.).
11. MOUNTFORT, G. 1973. *Tigers*, Newton Abbot, Devon.
12. SANKHALA, K. 1967. Breeding Behavior of the Tiger, *Panthera tigris*, in Rajasthan. *Inter. Zoo Yb.*, 7: 133–147.
13. SCHALLER, G. 1967. *The Deer and the Tiger: A Study of Wildlife in India*. Chicago.
14. SEIDENSTICKER, J., HORNOCKER, M., WILES, W., and MESSICK, J. 1973. Mountain Lion Social Organization in the Idaho Primitive Area. *Wildl. Monog.*, 35: 1–60.
15. SEIDENSTICKER, J., TAMANG, K., and GRAY, C. 1974. The Use of CI-744 to Immobilize Free-ranging Tigers and Leopards. *J. Zoo, Animal Med.*, 5(4): 22–25.
16. SHOREY, D. and EATON, R. 1974. Management and Behavior of Bengal Tigers under Semi-natural Conditions. In R. Eaton, ed., *The World's Cats, Volume II: Biology, Behavior and Management of Reproduction*, 204–221. Feline Research Group, Woodland Park Zoo, Seattle.
17. YOUNG, E. (ed.). 1973. *The Capture and Care of Wild Animals*. Capetown and Pretoria.

K. Ullas Karanth

MEL AND FIONA SUNQUIST

THE TIGER SINGLES SCENE

Early morning fog swirls above the surface of the water and mutes the colours of the landscape, turning everything to monochrome. From the tall simal trees moisture drips like rain on to the twenty-foot-high grasses below. A black-and-white pied kingfisher twitters, then hangs in the air over the grey water, wings blurring. Suddenly, a low, resonant roar resounds across the broad flood plain. It is an awesome sound that carries for about two miles. The tigress roars again and again, sixty-nine times in fifteen minutes, then stops; forest and grassland once more echo with the noisy squabbling of birds.

The roaring tigress was a young animal coming into her first estrus, and she was probably trying to inform the local male of her receptive state. A day later the male showed up and stayed close to her for forty-eight hours. These animals were part of the first long-term study of tigers and their prey, which was conducted in Nepal's Royal Chitawan National Park, where the Smithsonian Institution and the World Wildlife Fund collaborated with an enthusiastic host government. Largest of the world's cats, the tiger has historically epitomized the strength, cunning, and wild places of Asia. For thousands of years it has been the subject of myths, magic, and folk tales, but until recently little was known of this big predator's biology in its natural habitat.

Chitawan was selected as a study site because it has long been known to support a thriving population of tigers. In the 1930s it was the scene of lavish tiger hunts in which kings and dignitaries from all over the world participated. During a hunt in the winter of 1938/39, the prime minister of Nepal and his guests shot some 90 to 100 tigers in Chitawan Valley alone. At present, the park covers an area of 360 square miles, bounded on the north by the Rapti and Narayani rivers and on the south by the Indian border and the Siwalik Hills. Rivers fed by runoff from the high Himalayas divide the Siwaliks at intervals, forming the wide, fertile valleys known as duns. The mixture of grassland and riverine forest that covers the valley floors is maintained and constantly rearranged by the rain-swollen rivers. Tranquil for most of the year, the watercourses are transformed into silt-laden torrents during the three months of monsoon rains. Great chunks of forest are torn away from the banks and deposited elsewhere. The silt is quickly colonized by grasses, and within a year the open mudbanks are transformed into waving grassland.

It is a constantly changing, dynamic system. The rivers change course frequently, and often rapidly, leaving abandoned oxbow lakes that gradually become swamps and then dry land. These rich, marshy bodies of water become a focal point for a host of animals and birds. Ibises, open-billed storks and herons festoon the trees; rhinos wallow in the muddy water, and during the hottest months, tigers use the dense wet grass around the lakes as a resting place.

The mosaic of tall grassland and riverine forest on the flood plain is prime tiger habitat. There is plenty of water and excellent stalking cover, and the area supports a high density of ungulates. Major prey species include the elk-sized sambar and smaller ungulates such as spotted deer, hog deer, barking deer, and wild boar. A variety of other mammals are also found in the park, including the sloth bear, the leopard and the Indian ox, or gaur.

The study of a solitary, nocturnal and potentially dangerous predator such as the tiger presents enormous logistical problems, especially when visibility is limited by dense vegetation, as it is in Chitawan. To identify and follow individuals at a distance, we used radiotelemetry, which enabled us to monitor several animals simultaneously and provided information on how frequently they socialize and how they utilize space and resources.

As anticipated, direct observation of the tigers was rarely possible. Most of their activity occurred between late afternoon and midmorning and dense vegetation precluded attempts to make observations at rest sites during the day. However, we could follow their movements with the aid of radio signals and we tracked collared animals on foot, from elephant back or vehicles, and occasionally from aircraft.

Location data revealed that both male and female tigers maintained home ranges that were exclusive of others of their sex. Resident males occupied areas of about twenty-five to forty square miles and each male's range encompassed the smaller eight-square-mile ranges of three to four females. These home ranges are far too large to be maintained by physical presence alone, and the solitary tiger, depending solely on its own resources to catch and kill prey, cannot afford potentially damaging encounters. Instead, territories are maintained by visual and vocal systems and by a complex system of scent markings (deposited throughout an animal's home range and especially along the borders). Scent probably conveys information regarding individual identity, reproductive and social status, and the time the mark was made. These signals indicate that an area is occupied, but they do not necessarily keep other tigers out. Intruders do not remain in occupied areas for long, however.

We found that tigers moved around their territories frequently and were rarely found in the same place on two consecutive days unless they had made a large kill. They moved an average of six to twelve miles per night, with males moving significantly farther on a day-to-day basis. The tigers visited most parts of their home range every two to three weeks, and although these movements were related to finding and killing prey, it became obvious that they were also intricately linked to the maintenance of rights to an area. The maxim seemed to be 'use it or lose it', as on two occasions tigers had their home ranges appropriated by others when they failed to use all portions of their territory.

In the first instance of a home range loss, the resident male, which had occupied the territory for three years, ceased to visit its boundaries. He was found dead, probably poisoned by local farmers. Barely a month elapsed before the neighbouring male to the east began to make forays into the dead male's range, an area where he had never previously ventured. The neighbour eventually appropriated most of the dead male's range, effectively doubling the size of his territory from twenty-five to forty square miles, and increasing to seven the number of females with whom he could mate. He maintained this expanded home range until his death, visiting all portions of it with the same frequency as he had his smaller range; however, he spent less time in each place, probably because he had to move more often and travel farther to sustain his rights to the area.

The second range shift took place between a mother, whom we called Number One, and her daughter, the young roaring tigress. By the time she was eighteen months old, the young tigress was nutritionally independent of Number One. Although she continued to use her natal range, she made her own kills, and mother and daughter were rarely found hunting or resting together. In May 1975, when she was two years old and not yet sexually mature, the young tigress shifted her activities to a small island in the Narayani River, just outside her mother's home range. During the next nine months, she ventured into her mother's range on only nine occasions and all her travels were along the periphery.

In October, when she was two and a half years old, she came into her first oestros period. She started to roar periodically, and the resident male (her father) began to visit her on the island, an area where we had never previously located him. Indeed, in the previous nine months he had not been located within a half mile of her. The male's visits to the island coincided with the tigress's bouts of roaring, but any matings that may have taken place were not successful as she did not have cubs.

Her mother gave birth to a new litter in mid-December and she subsequently restricted her movements to a small area around the den site. When the cubs were a month old, the annual grass-cutting season began and thousands of villagers came into the park to harvest the tall grasses. Traditionally, the area is burned after the grasses are cut and, at the height of the fires in mid-February, Number One relocated her cubs, moving them out of the grassland to a far corner of her range (after charging the grasscutters in the area four times). She remained in the area of her litter and did not venture back into the grassland for some time. Within two weeks of Number One moving her litter, the young tigress was using a portion of her mother's home range. During the next six weeks she moved farther into the area. In March and April, Number One made a few excursions back into her former range, but it was too late; by May she had lost 90 per cent of it to her daughter.

During the time she was establishing her rights to the area, the young tigress spent a disproportionate amount of her time along its boundaries. In July, she successfully mated with her father and in October she gave birth to a litter of cubs in the same den site her mother had used. Meanwhile, Number One lost her litter of cubs and wandered around on the periphery of her old range for some time before establishing a new range adjacent to her old one. The circumstances associated with the range shift of these females appear to be unusual. Most female ranges remain stable and some tigresses in Chitawan have occupied essentially the same ranges for six years or longer. For instance, Number One's daughter is still a resident in the area she appropriated from her mother.

Male ranges seem to be less stable as the males compete for larger areas that include more females; female ranges are tied to the necessity of regularly and predictably securing food to raise young. Throughout the two years of our study, most females we observed were either pregnant or accompanied by dependent young, which remained with their mother until they were approximately twenty months old. Male cubs became independent earlier than their female siblings and wandered farther afield at an earlier age. When Number One's daughter was nutritionally independent of her mother but still using her mother's six-square-mile range, her male littermate was using a seventeen-square-mile area that included his mother's range and those of several other females. He was tolerated by his father, the resident male, and during the hot season they frequently rested along the same stream within 1600 feet of one another. Indeed, at this time the boundaries of his range corresponded almost exactly with the limits of his father's home range.

When he was thirty-four months old, he left his natal area and, although we made great efforts to find him, we never sighted him again. Three years later, a tiger with 'a strap around its neck' was sighted by local hunters in an area 150 miles away. It was almost certainly the same male, but we never managed to see him ourselves. To reach this area he would have had to travel through some very inhospitable country, and on several occasions cross one-and-a-half-mile-wide belts of agriculture.

Like most felids, tigers maintain a spatially and temporally dispersed social system where adults associate infrequently. They are, however, capable of socializing under special circumstances, as the hunting literature illustrates. Aggregations of six or seven tigers have been seen around bait kills, and there are reports of several tigers being driven out of the same small patch of cover where baits have been tied. In zoos, tigers can be maintained in pairs or social groups without problems, indicating that they can show a greater degree of social tolerance than might be anticipated from their normally solitary existence in the wild.

Apart from mother–young associations and occasions when animals were thought to be mating, we rarely found our study tigers together, either hunting or at kills. However, Charles McDougal, who observed tigers at bait sites at the other end of the same national park, but in areas where baits are regularly put out to attract tigers for tourist viewing, reports that aggregations are not uncommon. These aggregations are most likely at newly established bait sites and when baits are placed along home range boundaries. Even when two adults were present, he saw little aggression and animals seemed to feed quite peaceably, turn by turn, allowing first feeding rights to the individual that made the kill. He did notice that there were certain sex and age classes of tigers that used the bait sites more frequently; these included young males and adult females with large dependent cubs. Females with small cubs almost never visited the bait sites, probably because of the likelihood of encountering adult males, which have been known to kill cubs in these situations.

The most likely explanation for the differences observed in the same park is that tigers rarely have the opportunity to socialize at natural kills. We found that tigers made a large kill roughly once every eight days and remained nearby, alternately feeding and resting until the carcass was consumed. Given the nature of the tiger's social system, and the dispersal of kills throughout an animal's home range, we were not surprised that two or more adults were rarely found feeding together. The chances of one tiger encountering another on a natural kill are low. Perhaps the most surprising observation is that there is so little aggression in groups feeding at bait sites, where tigers have been seen to behave as socially as lions.

The solitary tiger and the social lion are strikingly similar morphologically. Their gestation period, litter size, tooth-replacement patterns, and age at sexual maturity are nearly identical, and the skulls and dentition patterns of the two species are almost indistinguishable. Both show a similar degree of sex differences, and both are physically adapted for the single-handed capture and killing of prey as large or larger than themselves. And if large brain size is an index of social capacity, then the tiger should be as social as the lion. Why then isn't it?

The evolutionary histories of the two species are quite different. The initial appearance of the lion apparently came after the explosive Pliocene radiation of the artiodactyls (hoofed mammals) into the grasslands, while tigers probably evolved in response to the evolution of forest–grassland-edge ungulates in Southeast Asia. The lion, living in more open habitat where prey are abundant and form large herds, has evolved a complex social unit based on extended mother–daughter relationships. The demands upon the tiger of living in dense cover, where relatively small prey are scattered and time-consuming to find, have not fostered the development of social groups. Under these circumstances there is no selective advantage to cooperative hunting or feeding, and it is more efficient to hunt alone and thereby reduce interference from others.

Despite differences in hunting strategies and social structures between lions and tigers, both species exhibit similarly flexible responses to variations in prey density and dispersion. For example, lions living in the harsh conditions of the Kalahari Desert, where the habitat cannot support a dense population of large herbivores, are usually solitary or form small prides of two to five animals. Their diet consists mainly of small or young mammals, and home ranges are large and overlap extensively. This is in direct contrast to the lions of the Serengeti Plains. There prides are large, and the lions maintain small, exclusive home ranges. In this area there is a dense population of large prey, and small animals form an insignificant part of the lions' diet.

In the Soviet Far East, where prey are scarce and make large seasonal movements, tigers maintain extremely large home ranges. Males may have ranges up to 400 square miles, and three or more females share overlapping ranges of 40 to 150 square miles. Defense of food resources under these conditions would seem to be uneconomical, whereas in Chitawan, females are able to meet their resource requirements within a much smaller area, and maintenance of exclusive ranges becomes feasible.

The capacity for tigers to socialize at steady food sources with little overt aggression raises interesting implications for the evolution of some type of sociality. Combined with the possibility of home range inheritance by female offspring, it suggests that two or three females sharing the same home range are related. As areas available for reserves dwindle in size and become more like islands, the tiger's capacity to socialize may be an important factor in its survival. A given area cannot support an unlimited number of tigers, however, and as national parks become surrounded by agriculture and dispersal corridors are eliminated, tigers will be unable to enter or leave the preserves. The consequences of this development will probably be a disruption of the tiger's social system and increased conflict with farmers and their livestock. Thus, the tiger's future is uncertain at best. Proper long-term management decisions must be made now if these big cats are to survive.

K. ULLAS KARANTH

UNDERSTANDING TIGERS

Perched fifteen feet up in the fork of a *Randia* tree, I wait with dart gun in hand. Sunlight filtering through the leafy canopy creates a harlequin pattern of colour, light and shade in the thick brush below. I can hear muted curses as park ranger Chinnappa directs my team of trained trackers riding atop elephants to push their way through dense bamboo. Somewhere in the space between us is a tiger sleeping off a meal from a buffalo he killed the previous night. I am hoping that the cat will be disturbed by the elephants, slink away from them, and head towards me. If he does, he will face a three-foot-high barrier of white cloth strung taut across the bushes. Although the 500-pound, lethally armed predator could effortlessly rip through this flimsy stockade, the stark white cloth against the green jungle will make him wary, and he will most likely search for a way out. If the plan works, the tiger will eventually emerge through a 50-foot opening in the quarter-mile-long funnel beneath my perch, offering me a chance to shoot him with a tranquilizing dart.

Listening for animal alarm calls betraying the tiger's stealthy passage, I muse about the ingenuity of the Asian shikaris, or hunt assistants, who invented this technique, once employed by royalty to slaughter tigers. Ironically, it is now one of the tools I use to help to preserve these big cats. In order to answer some basic questions about the species (How many can live in a particular forest? What kind of prey do they eat? How long do tigers live? And, above all, how are tigers responding to man-made changes in their environments?), I must first be able to follow the cats as they are born, learn hunting skills, disperse, establish territories, find mates, produce offspring, and finally die. To do that, I have had to adopt a new skill: radiotelemetry.

Which is why I am on a tree in Nagarahole wildlife reserve in southern India, straining every nerve, listening and peering, hoping for a glimpse of amber shadow. Minutes earlier, I had heard the tiger's deep-throated growl of annoyance as the beaters on elephant-back forced him to abandon his kill in a narrow gully. There has been no sign of the tiger since then, not even the cackling of the silver-hackled jungle fowl that usually signals a tiger's approach. Did he turn back and sneak away between the elephants? Worse, would he bolt down the trail under my tree, ruining any chance of safely darting him?

My skills and training, gained under the patient tutelage of Melvin Sunquist, a renowned expert on big cats, were being put to the test. Suddenly, in the dense cover I see a ghostly shadow move. Adrenalin races in pounding pulses. Then I spot him: a brief glow of gold as he calmly pads down the trail, massive head swaying from side to side, muscular body a picture of power and grace. I slowly swing my dart gun around, hoping his razor-sharp vision will not catch the movement. My best chance is to get him broadside, through an opening in the cover about twenty feet away. As his head, shoulders, and flanks appear in the cross hairs, I squeeze the trigger. There is a soft *plop* as the gun shoots a red-tailed syringe, which buries itself in the tiger's massive thigh. He growls, stops, looks around. Damn these stinging bees, he seems to say. I hold my breath. Suspecting nothing, he continues down the trail and through the gap. As he strides out of sight, I whisper into my walkie-talkie, 'We got him!'

Fifteen minutes later, we find him fully sedated under a tree. Because he is so handsomely striped and well proportioned, we call him Mara, a local Kannada name for the god of love. Soon, broadcasting beeps from a transmitter affixed around his neck, Mara joins three other tigers I have already radio-collared in Nagarahole.

My daily routine begins at dawn, when I drive up Kuntur Tittu, a hill at the centre of my study area, to listen to signals from my tigers. All through the day, and sometimes the night, I drive, walk, or ride elephants, criss-crossing the park to keep track of the tigers as they move miles silently in cover, searching for prey. Sometimes, when they are looking for mates, their deep roars reverberate across the landscape. I have seen tigers stalking, mating, and chasing leopards up trees, but during most of the day they sleep. Tracking tigers is not all fun and excitement; often it is about as thrilling as land surveying—painstaking map-and-compass work. But there are rewards for entering the secret world of tigers.

I started my tiger project in the 250-square-mile Nagarahole reserve in 1986, supported by the Wildlife Conservation Society and the US Fish and Wildlife Service. I hoped to build on earlier studies by George Schaller in central India and by biologists of the Smithsonian Institution in Nepal. They had learned that adult females form the core of tiger society, defending exclusive territories from which other breeding females are excluded but in which subadult offspring are tolerated. These tigresses mate with a territorial male who usually has a larger range overlapping those of three females. Young tigers leave their mothers at about two years of age and try to establish their own territories.

When I began my project, little was known about how many prey animals lived in the forests. So, I cut several two-mile-long trails through Nagarahole and repeatedly walked them, counting prey species. After walking 300 miles and plugging the resulting data into a computer program for analysis, I discovered that Nagarahole forests are packed with prey at a density of 170 animals per square mile. Because the reserve is so rich ecologically and vigorously protected by Chinnappa and his dedicated staff, Nagarahole tigers can choose from an astonishing array of food items, ranging from spaniel-sized muntjac deer to wild cattle called gaur that can weigh more than a ton.

To determine what the tigers kill, my trackers and I collected scats—a deceptively neat name for the smelly excreta deposited by carnivores—and looked for kills. After examining hair and bone fragments in 490 scats and looking at 154 kills, I learned that in Nagarahole, unlike other areas where they prey chiefly on deer, tigers routinely kill adult gaur five times their own weight.

In addition, radio-tracking revealed that the cats hunt primarily between dusk and dawn, carefully searching in broad, zigzag sweeps. They are solitary stalkers, ambushing prey from a hiding place. The tremendous initial impact of the cat brings the prey down, and a swift grip on the throat strangles it. I found that prey species were being heavily cropped by tigers, leopards and wild dogs. Roughly a third of the deer, pig and gaur in Nagarahole were younger than two years of age. Yet, their populations held steady, and tiger numbers were at their highest densities compared to tiger populations in other regions. And there was hardly any livestock evident in the tigers' diet.

We also discovered that Nagarahole tigers have very high death rates resulting from fights for space, kills and mates. Fewer than half the cubs grow to adulthood. From among these survivors, only about half make it through the fierce competition to reach breeding age. Because females breed at the early age of four years and produce three to four offspring every third year, however, tiger populations could continue to grow if protected from humans. Even relatively small tropical reserves could maintain thriving tiger populations. Sad to say, this is not the case in most of the tigers' range. Nagarahole has ten times the density of cats found in forests only a few miles away.

My studies in Nagarahole suggest that the single most important reason that tigers are scarce over much of Asia is the loss of their prey base. Every day, in fact, thousands of villagers enter forests around their homes to shoot, snare, and trap the tigers' favoured food. Except in a few well-protected sanctuaries, such uncontrolled hunting for the pot by local people has driven prey populations down to levels unable to support tigers.

From my data, I estimate that a tiger kills once every week or so, taking about fifty animals a year. A mother tiger, however, has to feed her young as well; she needs to take down roughly seventy prey animals a year, most within a short radius of her voracious litter. To allow an annual 'crop' of prey that will support one tiger, the prey base must number about 500 animals. Therefore, for every fifty deer and pigs killed by villagers in a year, there is prey for one less tiger.

In many places, surviving cats are forced to turn to livestock, drawing swift reprisals from people who shoot, snare and poison the cats, and even burn the forests to kill or drive them away.

Twenty-five years ago, efforts were launched worldwide to save tigers. Most Asian countries harbouring wild tigers passed laws to control tiger hunts by sportsmen and establish wildlife sanctuaries in which human intrusions such as farming, grazing, setting fires and timbering were forbidden. *Implementing* these laws was another matter, however. The burden fell on underpaid forestry services personnel who often did not possess the necessary resources, training and political backing. Consequently, the tiger's overall decline continued, except in parts of India and Nepal, where half the world's tigers live. There, thanks to disciplined efforts of wildlife managers, backed by political will at the highest levels, several fine reserves were established. At least in these few pockets, tiger populations recovered.

Through the 1980s, while India's reserve managers continued to report dramatic countrywide increases in tiger numbers, IUCN (World Conservation Union) and other conservation agencies were eager for a 'success story.' Ignoring the warnings of some biologists, they repeated these numbers without hard data to back them up. Recently, when evidence of the horrendous scale of tiger poaching in Asia surfaced, their complacency was shattered. It became clear that the tiger numbers being trumpeted by various Asian governments were practically worthless.

Traditionally, wildlife managers in most Asian countries have guessed at the tiger numbers in their reserves. In Russia, managers scout the countryside after a fresh snowfall, assuming that widely separated sets of tracks belong to different tigers. Indian wildlife managers, untrained in population monitoring methods, make the demonstrably untenable claim that they individually identify every tiger in that vast country (one-third the size of the United States) simply by looking at tiger paw prints lifted from dusty trails. In the absence of reliable monitoring, there is no way of knowing whether efforts to save tigers are succeeding or failing.

As the international conservation community responds with increasing urgency to the tiger crisis, there is a critical need for Asian wildlife managers to use objective, rigorous methods to measure the effectiveness of protective efforts. If not, these efforts are bound to flounder, much like a business enterprise that runs without ever drawing up a balance sheet. Therefore, my research focuses on developing better techniques for counting tigers and for training park managers and biologists in these methods.

At the simplest level, tiger monitoring involves repeatedly walking the forest trails and roads that cats use regularly. Tigers mark their passage by scraping the ground and depositing scats—their 'calling cards' for communicating with other tigers. The frequency with which we encounter such signs can yield precise indicators for tracking changes in cat numbers over time. Although these indices do not reveal actual numbers, they do show whether the numbers are declining or increasing.

Because tigers are secretive, nocturnal and scarce, visually counting them is impossible. Tiger numbers are correlated to the numbers of deer, pigs, wild cattle, and livestock on which they survive. Tigers annually kill about 8 to 10 per cent of available prey, so realistic estimates of tiger populations can be derived from accurate prey counts.

In high priority areas, such as special tiger reserves, it may be necessary to estimate population densities of tigers by identifying individuals. But because the paw prints of the same tiger may look different due to variations in soil type, slope, movement and other factors, prints are not reliable. Shape and arrangement of stripes on tigers *are* distinct enough, however, for even schoolchildren to identify individuals. It is almost as if the tigers have their names written on their bodies. If clear photographs can be obtained, the animals can be counted accurately.

In Nagarahole, I used commercially available camera traps to count tigers. A camera trap consists of a flash-equipped, autofocus camera wired to an electronic tripping device not unlike those used in burglar alarms. Every evening, I would go to a part of the forest and set the traps across trails frequently used by tigers. While I slept peacefully in camp, the tigers would pad down the trails and, after breaking the invisible electronic beam that triggers the camera, take their own pictures. The resulting portraits helped me to gradually identify their coats. With the help of statistical analyses, these identifications showed that my study area was packed at a density of forty tigers per 100 square miles. These numbers tallied perfectly with the tiger densities I had estimated based on prey availability.

Only a hundred years ago, tigers ranged from Siberia's temperate woodlands to Iran's reed beds, from India's deciduous jungles to Bali's rain forests. Today, they live precariously in a few patches, having retreated in the face of relentless pressures—clearing of land for farming and grazing, woodcutting for fuel and timber, overhunting of prey and hunting for the illegal trade in tiger parts. Tigers are truly safe only in some well-protected reserves in India and Nepal. Elsewhere, the future of these magnificent felines is uncertain. The fragile ecological webs that bind tigers to other living creatures have been seriously ruptured.

Saving the tiger involves making difficult decisions, decisions we have been putting off for twenty years. We must immediately institute strict protective measures inside wildlife reserves for tigers, their prey, and their habitats, and we must crack down on traffickers outside reserves. It also means relocating forest-dwelling peoples in a humane fashion and abolishing timber and other forest product exploitation from critical tiger habitats.

In Asian cultures the tiger is a magical symbol, epitomizing power, splendour and ability. It is also a valuable icon for modern mega corporations selling profitable products in global markets. Above all, the tiger symbolizes, perhaps more than any other animal, mankind's struggle to protect at least a part of the natural world we share with our fellow creatures. If we act rationally and deploy our resources wisely, there is still time to save the tiger.

JOHN GOODRICH, DALE MIQUELLE, LINDA KERLEY AND EVGENY SMIRNOV

TIME FOR TIGERS

December 15, 1998[1]*—'Today I carried Nadia off of a hill in pieces; her skin, her legs, her spine and ribs, her head. All were scattered across a ridge-top; her hide buried here in the snow, her head several yards away, chopped from her neck and carelessly tossed down the hillside, her body above on the ridge, lying in packed snow stained with blood and excrement. Everywhere were the tracks of her cubs, wandering hungry and cold. Nadia was a radio-collared tigress who had been shot by Vasily Mulucov, a poacher from Plastun. Tracks in the snow told the story. Less than a kilometre from where I stood, the poacher left his WAZ (a Russian jeep) on the road and began tracking a wild boar he had wounded. Seventy-five metres from where I stood, the poacher came across the tracks of Nadia and her three cubs; these tracks were on top of the boar tracks. Whether he wanted to kill a tiger or was just too stupid to realize moving forward would be dangerous is known only to him, but he did move forward and as he came over the ridge-top on which I was standing, he surprised Nadia. She charged aggressively to within fifteen metres and he shot her . . .*

As I stood on the ridge-top looking at Nadia's remains, I remembered it was three years ago almost to the day that my Russian colleagues and I crept through the moonlit forest, snow crunching under our feet, hearts pounding, breath coming out in great clouds, the air so cold and quiet that every sound seemed as though it rattled through the forest like an earthquake. Then there was a low growl like distant thunder—as much felt as heard—a warning, an approaching storm. My adrenaline jumped and every hair on my body stood on end as a shiver rippled my skin from head to toe. With every step, the growls became louder, more urgent, and finally became roars. We heard her lunge in the darkness, green eyes flashing in the beams of our flashlights. Then, a soft 'pop' as a tranquilizing dart flew through the air and found its mark. Ten minutes later, she was sleeping like a baby as we removed the foot snare in which we had caught her and fitted a radio collar around her neck. Within four hours, she was moving off upstream, perhaps a bit hung over, but none the worse for wear.'

We lost an important study animal, but Nadia's death was not in vain. With this incident, we began to see a pattern that might help put an end to such poaching incidents. Nadia, like so many other tigers, was killed because a road provided access for a poacher. It is clear that where there are roads, tigers get poached or hit and killed by cars. Where there are no roads, tigers can live to a ripe old age and die natural deaths.

A simple but unrealistic answer: Keep people out of the forests or keep tigers only in protected areas. The more complicated but necessary solution must answer this question: How do we ensure that local people retain access to forest areas to collect wood, meat, berries, and fish, yet ensure some level of security for tigers and other wildlife? This is a difficult question, but we think we have some of the answers.

Nadia is one of more than thirty tigers we have captured and fitted with radio collars during the past ten years as part of a Russian–American project between the Wildlife Conservation Society's Hornocker Wildlife Institute and the Sikhote-Alin State Biosphere Zapovednik. The goal is to study tiger ecology and to determine ways in which people and tigers can coexist, while meeting the needs of both.

After Nadia was outfitted with a radio collar, we followed her every move for three years. We knew what and how much she ate, the size (240 square miles) and boundaries of her territory, with whom she mated, and when and where she gave birth to her three cubs. We have collected similar information on the other cats, including Olga, who is our first collared tiger (caught in February 1992) and still out there today (see From the Field, page 6). Among our findings is the relationship between roads and tiger mortality.

We studied the fates of radio-collared tigresses and their cubs that prowled three types of territories: remote areas in the Sikhote-Alin Zapovednik with no roads, areas surrounding the Zapovednik with secondary roads (not regularly maintained, but allowing public access into forested lands), and areas with primary roads (maintained year round and providing access between towns or villages). Primary roads are paved or hard-packed dirt and allow traffic to move at high speeds, while secondary roads are suitable only for 4-wheel drive vehicles for part or all of the year.

We monitored the roads from 1992 to 2000 and followed ten tigresses and thirty-seven of their cubs from fifteen litters. Annual adult female survivorship was 100 per cent in the areas without roads. In areas with secondary roads, survival was 89 per cent, and with primary roads, survival was a mere 55 per cent. We found a similar pattern for cubs: 90 per cent survived in roadless areas, while 40 per cent survived in areas with primary roads. All of the tigresses that died were poached, and most of the cubs died because their mothers had been poached and they were too young to survive on their own. Two cubs were hit and killed by cars, as was one radio-collared adult.

Perhaps even more disturbing is the pattern in the region where the road connecting the towns of Terney and Plastun cuts across the Zapovednik. Although it receives much less traffic than many other roads in tiger habitat, it is the most heavily travelled of all the roads in our study site. Because ungulate populations are very high here, tiger reproduction should be high and many young should be successfully raised to disperse into neighbouring areas. We have monitored five radio-collared tigresses here since June 1992. Lena, our first tigress in this area, had had four cubs in 1991, but all of them died—three of unknown causes and one in a car accident. A month after her capture, Lena gave birth to four more cubs. Then, in November 1992, she was shot by a poacher on the road along Koonalayka Creek. We were able to capture her four cubs. Two of them died from congenital defects, and we sent the others to the

Omaha Zoo in the United States. We made this decision because these cubs would have died without human care and so were already lost from the population.

In good tiger habitat, a territory does not remain vacant for very long. Lena was quickly replaced by Katya, whom we also captured and radio collared. In November 1995, Katya gave birth to one cub, and successfully raised the youngster until it was twenty-two months old and ready to wander off to find its own territory. Not long afterward, in October 1997, this cub was hit and killed by a truck on Khaunta-mi Pass. In the meantime, Katya had given birth to a new litter, in July 1997, but she was poached in October and we never found those cubs. At three and a half months, they were too young to survive on their own and surely died.

Katya was soon replaced by two tigresses: Natasha, who had been Katya's neighbour to the west, and Natasha's two-year-old daughter, Emma. We had captured Natasha in 1992 and had already radio-tracked her for five years. During this time she held a territory in a roadless area. Natasha took advantage of the vacancy created by Katya's death to move to a territory with higher prey densities. She used the southern half of Katya's former territory, and her daughter Emma used the northern half.

Emma didn't last long. She was poached in the summer of 1998. In December of that year, Natasha gave birth to a litter of four on Camel Mountain. One of these cubs was killed by a small predator within two weeks, and a second later died of unknown causes. In July 1999, Natasha was shot by a poacher in the same area as Lena had been shot. Natasha's two remaining cubs were just seven and a half months old, but they survived. We captured one, Alec, in September and fitted him with a radio collar. His sister thwarted our efforts, but we know she survived until she was at least ten and a half months of age. Alec dispersed to the south. His transmitter ceased to function when he was two years old, most likely because he was shot and the collar destroyed.

Meanwhile, Natasha was replaced by Lidya, whom we captured in Koonalayka in October 1999. The following May, Lidya gave birth, but the cubs died of unknown causes a few months later. Lidya gave birth again this past summer and is still alive today.

To sum up, four of five tigresses in this region were poached, roughly one every two and a half years. Even more important, of 12 cubs born in the area, we know that ten died. Only Alec and possibly his sister survived to disperse to new territories. Remember, this is excellent tiger habitat and should be producing surplus tigers to populate other areas.

In contrast, let's look at Mary Ivanna and again at Natasha. Before she moved to take over Katya's territory, Natasha used a roadless area for five years and produced two litters, one with two cubs and the second with three. All survived to independence. Mary Ivanna gave birth to two litters totalling five cubs, four of which dispersed. In 1995, Mary Ivanna, like Natasha, moved to a new territory containing a primary road. In less than two years she was poached. While they were living in areas without roads, Natasha and Mary Ivanna and 90 per cent of their cubs survived, but not long after moving to areas with primary roads, both tigresses were poached.

We have followed several radio-collared male tigers as well, but all of them lived in territories bisected by primary roads, so we can't make the same comparisons about male survivorship. Nonetheless, we can look at male mortality patterns in relation to roads. Alexei was hit by a car near Kaimenka in December 1998. Zhenya, whose territory was in the same area as those of Lena and Katya, was poached in April 1999. Igor and Maurice disappeared; that is, their transmitters stopped functioning and from winter track surveys, we knew they no longer inhabited their former territories. These two tigers were almost certainly poached and their collars destroyed. Both disappeared while travelling less than half a mile from primary roads.

We are trying to reduce poaching. International groups have funded a programme called Inspection Tiger, a Russian governmental organization working to stop poaching and alleviate tiger–human conflicts. In addition, many of the larger zapovedniks have received foreign monies to maintain their own anti-poaching teams. Russian customs officials are working hard to stop exports of tiger and other wild animal parts, and groups such as the Wildlife Conservation Society are trying to reduce demand for tiger parts in China and other countries. Despite all these efforts, poaching continues.

We need to do more. If we can make it more difficult for poachers to get access to some areas, we can increase both adult and cub survival. Roads that are not necessary, especially those created for logging and other natural resource extraction, could be destroyed or made impassable. When a logging operation is finished, the roads remain and provide easy access for lazy poachers and a quick way for them to get the meat out. In Russia, legal hunters normally have a set of cabins, and they will come and go with or without roads. This approach would not work in all cases, because some roads are too important to local people. But if we can carefully close down those that no longer serve a useful purpose, we can greatly increase survival of the prey species upon which both hunters and tigers depend.

On some important secondary roads, it may be possible to set up gates manned with guards whose job it is to limit access to those with permission and a reason to use the area. This option gives law-abiding citizens opportunities for fishing, wood-collecting, and berry picking. And because the guards can ensure that guns are not illegally brought on to the land, it can largely eliminate poaching. Poachers willing to walk long distances will still have access, but in our experience, only a small percentage of poachers will make that effort.

The concept of road closures is relatively new in the Russian Far East, but it has been successful in North America and Europe. There, as in Russia, people will tear down blockades, find ways around road closures, and expend large amounts of effort to retain access to their favourite poaching areas. Persistence in rebuilding gates and providing personnel to enforce closures are key to long-term success.

Locations of new roads should be carefully planned to avoid sensitive areas. Often, roads are constructed with no review of potential environmental impacts. Unless there is a way to control where and how roads are built, as road construction continues and access increases, there soon will be no wild places left in the Russian Far East.

The only workable tiger conservation plans will be those that improve conditions for both tigers and people. After all, for tigers to exist, people have to want them to exist. We believe that controlling road use is a win-win situation and will benefit everyone.

* * *

In North America and parts of Europe, closing roads to protect valuable wildlife resources is an accepted practice. In Russia, however, this is still a new concept. To demonstrate the effectiveness of road closures, we launched three 'demonstration' road closures to show local hunters and wildlife biologists their value.

Since 1998, the Siberian Tiger Project has been working with Vladimir Valeechko and the Terney Hunter and Fisherman's Society—as well as with a non-governmental group called Terney, Taiga, Tiger—to close two important drainages within the Terney hunting lease. We closed the road to upper Sheptoon (Mala Seenansha) River, where Nadia was poached, and with a bulldozer, rendered the road into upper Beriozavey Creek, where natural salt licks concentrate deer and other ungulates, impassable. During 2000, an unknown party used a bulldozer to reopen the Beriozavey road, so we had to 'deconstruct' it again. Valeeckho, initially a sceptic, now says that these closed basins are the only places where ungulate populations have thrived in his hunting lease.

Southwest Primorski Krai is key habitat not only for the Siberian tiger, but also for the even more endangered Far Eastern leopard, whose population numbers fewer than 50. The Wildlife Conservation Society signed an agreement to support the construction of roadblocks and guard cabins on the three primary drainages that provide access to Nezhinskoe Hunting Lease. This area, within easy driving distance of Vladivostok and Ussurisk, has incurred heavy wildlife losses to poaching. Here, we are working with V. Vasiliev of the Naval Hunting Society and V.V. Aramilev of the Institute for Sustainable Use of Natural Resources. With adequate support, Nezhinskoe Hunting Lease can be a model of how to control poaching, provide good hunting opportunities for licensed hunters, and provide adequate prey densities for both hunters and large carnivores.

The Tavisa region, just north of Terney, is considered a key wintering area for ungulates. As recently as ten years ago, one could see herds of red deer on the winter hayfields and large groups of wild boar in the oak forests. With such an abundance of prey, it is not surprising that Tavisa was a favourite hunting ground for Olga, the first tiger collared by the Siberian Tiger Project. However, with construction of a new road providing easy access from Terney, and almost no control on hunting, ungulates have been virtually eliminated from this region. Olga[2] still lives in the general vicinity, but concentrates her activities to the north or south of Tavisa. WCS has joined forces with the county administration and the leader of the local federal anti-poaching patrol 'Inspection Tiger', B.I. Litvinov, to reverse the trends in this area. The plan calls for limiting access to Tavisa by building a guard station and a gate. People owning farms in the region, fishermen, and people harvesting hay will be allowed in. Those without a valid reason to enter the area will be turned back. We will institute a programme to monitor prey numbers to determine if, in fact, the road closure results in increases in ungulate numbers. It is our hope that, in the not too distant future, we will once again see herds of red deer wintering in the fields of Tavisa, and that Olga, or perhaps one of her offspring, will again consider Tavisa a favorite hunting spot.

Editor's Notes

1. This is an entry from John Goodrich's field diary.
2. Tragically, tigress Olga was killed by poachers after this article was written.

K. Ullas Karanth

GEOFFREY C. WARD

MAKING ROOM FOR WILD TIGERS

I first saw the tigress called Sita in 1986 in Bandhavgarh National Park in the Central Indian state of Madhya Pradesh. She lay asleep on a hillside when the elephant I was sharing with the Indian naturalist Hashim Tyabji found her, full-bellied after eating from the spotted deer that lay next to her. She was exhausted by the steady strain of having to feed and care for her first litter of three cubs, whose mewing I could just make out from still higher up the slope amid the expectant cawing of crows that teetered in the branches of the surrounding trees.

We were just thirty feet or so from the tigress, close enough to hear her steady, sonorous breathing. Over the next half hour or so three more elephants bearing tourists came and went. Cameras clicked. One over-eager photographer dropped a canister of film, and the mahout loudly ordered his mount to pick it up with its trunk and return it to its owner. The tigress slept on, oblivious. Nothing seemed to faze her—nothing, until a rufous-and-white tree pie smaller than a North American magpie fluttered down onto her kill. She was up and fully awake in a millisecond, swatting at the terrified bird with one enormous forepaw and roaring so loudly the sky seemed to split.

It was the first time I'd been so close to so much tiger anger. I was thrilled but frightened too, and looked to Hashim for an explanation. Why had she become so furious so fast?

He smiled. 'Tigers', he told me, 'do not like to share.'

Last spring I was back at Bandhavgarh starting out again at dawn, riding the same elephant with the same mahout, and looking for the same tigress.

Sita is nearly sixteen now, unusually old for a tigress in the wild, and she has given birth to eighteen tigers over the eleven intervening years. Just seven made it to adulthood. The rest died or disappeared: one young male was killed by an adult male seeking to displace the cub's father, and his sister drowned in a monsoon flood; a female cub struggled with physical deformities, then seemed to waste away. And all three offspring from Sita's fifth litter—born in March 1996—died two months later. No one knows for sure what happened to them.

Still, less than ten days after the loss of her fifth set of cubs, Sita was seen mating again, with the big, testy resident male. He is nicknamed Charger because of his enthusiasm for doing just that, once clawing his way up the back end of a tourist elephant when it got too close and in the process traumatizing the visitors on board.

The forest was still dark as we set out, and our fleet of five elephants moved along in almost total silence. But around us there were already morning sounds: peacocks called from their nighttime roosts and were answered by the raucous crowing of jungle fowl, gaudy ancestors of the domestic chicken. Grey langur monkeys gave out the low self-satisfied hooting with which they greet the day and warn one another to be on alert.

Ahead of us we could just make out a broad stretch of swampy grassland, where I hoped to see Sita again and, if I was very lucky, catch a glimpse of the new litter of cubs said to be somewhere in the area as well.

One evening not long before I left for India, the former head of one of America's best known zoos appeared on a nationally televised talk show, clad in safari clothes and holding a live cub in his arms. Tigers kill their prey by breaking its neck, he assured his host, then bury it overnight. Only 2000 tigers remain in the wild and all of them will be gone by the year 2000, he suggested. The only surviving tigers will be in zoos.

His heart may have been in the right place, but everything he said was suspect. In fact, tigers throttle most of their prey—in the field the distinctive throat wounds left by their big canines often provide the best evidence that they have been at work. And while they do conceal their kills as best they can in brush or leaves or tall grass, they do not bury them. Nor has anyone any real idea how many tigers survive in the wild—the dearth of reliable numbers is one of the most frustrating aspects of tiger conservation. There is no question that the species is in trouble throughout its remaining range, but there is no reason to suppose tigers will all be gone by the turn of the century—or anytime soon after that—provided governments intensify their efforts to protect them, good science is applied to their conservation, and well-meaning alarmists don't convince the public that their rescue is a lost cause.

Tigers are generally believed to have evolved in southern China more than a million years ago and then to have prowled westward towards the Caspian Sea, north to the snow-filled evergreen and oak forests of Siberia, and south, across Indochina and Indonesia, all the way to the lush tropical forests of Bali. Their modern history is admittedly dispiriting. Into the 1940s eight supposed subspecies persisted in the wild. Since then, however, the tigers of Bali, the Caspian region, and Java have vanished, and the South China tiger, hunted as vermin during the regime of Mao Zedong, seems poised to follow them into extinction; fewer than 30 individuals may now survive outside zoos, scattered among four disconnected patches of mountain forest, probably too few and far between to maintain a viable population ever again.

Just four other subspecies remain, and well over half the tigers in the wild are believed to live in India and neighbouring Nepal and Bangladesh. When I started writing about the tigers of India in the early 1980s, their future, at least, seemed assured. Shooting them had been banned since 1970, and there were stiff penalties for anyone caught trying. Project Tiger, undertaken at the instigation of Prime Minister Indira Gandhi in 1973, had set aside nine national parks for special protection (fourteen more have been added since). The core areas of these tiger reserves, off-limits to humans, were meant to be 'breeding nuclei, from which surplus animals would emigrate to adjacent forests'. Broad buffer zones, into which human incursions were limited, were to protect the breeding grounds. It was an extraordinary commitment for a relatively new nation beset by other, more pressing challenges; no Western country has ever mounted so serious an effort to save a magnificent but potentially deadly predator in such proximity to its citizens.

And it seemed to be working. In 1984 forestry officials declared that the number of wild tigers in India had more than doubled, from 1827 individuals to better than 4000. Project Tiger seemed so successful, its reserves were said to be so full of tigers, that some conservationists worried about what would happen to all the surplus animals.

Then came the bad news. The assassination of Mrs Gandhi in 1984 swept from the scene Indian wildlife's most powerful defender. Afterwards, as effective political power slowly shifted from the central government at New Delhi to local politicians in the individual states, enthusiasm for defending India's jungles slackened under pressure from ever growing numbers of poor voters who saw them primarily as easy sources of fuel and fodder. The authenticity of the gains Project Tiger had claimed came into question as well. No one doubts that the number of tigers really had risen. But in reaching their ever more impressive tallies, forest officials had relied on identifying individual pugmarks, or paw prints—a system since shown to be inexact—and then, concerned for their jobs if the number of animals under their care didn't steadily climb, some had inflated their findings well beyond the numbers the resident prey base could conceivably have sustained.

Meanwhile, the human population rose. Promised corridors were turned into farmers' fields, inundated by dams, and honeycombed with coal mines. Many of the 'adjacent forests' to which the corridors were meant to lead simply vanished. There were fewer and fewer places to which young tigers could disperse and more and more conflicts between tigers and human beings.

Then, beginning about 1986, something else began to happen, something mysterious and deadly. Tigers began to disappear. It was eventually discovered that they were being poisoned and shot and snared so that their bones and other body parts could be smuggled out of India to supply the manufacturers of Chinese traditional medicines. After the virtual disappearance of the South China tiger in the late 1980s, stockpiles of tiger bones were depleted, and resupplying them became a big business. No one knows how many tigers in India fell victim to this illicit trade, but the figures compiled by Ashok Kumar and Belinda Wright, whose tiny Wildlife Protection Society of India has spearheaded the fight against poaching on the subcontinent, represent only a small part of a very grim picture—94 tigers killed in 1994, 116 in 1995. Most poaching goes undetected, so the real number of butchered tigers must have been much higher.

Ranthambhore National Park in eastern Rajasthan—with its blue lakes, sprawling hilltop fortress, and ancient ruins scattered through the forest—was the pride of Project Tiger when I first visited it, famous both for the number of its predators and the astonishing ease with which they could be seen and photographed. On one especially memorable day at Ranthambhore in 1986, I was privileged to watch nine different tigers hunting, courting, caring for their young. My guide was the park's long-time field director, Fateh Singh Rathore. It had taken him eighteen years of unremitting work to turn a small and largely lifeless former hunting preserve into perhaps the most celebrated national park in India, and he had almost lost his life in the process: angry herdsmen beat him into insensibility, determined to graze their buffalo in grasslands he was no less determined to hold in reserve for the deer and wild boar on which his tigers fed.

He has not been in charge of the park for a decade now, but he still lives on its western edge, and it was he who first noticed that the tigers he had observed for years were vanishing. By 1993, he believes, perhaps as many as twenty had been killed—nearly half of those then thought to be living in the park—so catastrophic a loss that even the tiger's extraordinary ability to rebuild its population may never be able to compensate for it.

A sign near the railroad station at Sawai Madhopur proudly welcomed me last spring to the 'City of Tigers', and the road that leads out of town towards Ranthambhore was lined with hotels that provide accommodation for thousands of eager wildlife viewers who still come here each year from all over the world.

Ranthambhore has always been an embattled island. It is small—just 150 square miles—and surrounded by ever growing numbers of desperately poor people with little time or sympathy for tigers. The area once meant to be a buffer zone had been stripped bare by foraging livestock. And without strong backing from the state government of Rajasthan, forest officials no longer seemed willing to risk their lives resisting encroachment by human beings or their ravenous herds.

Nongovernment organizations, most notably the Ranthambhore Foundation, have laboured hard to persuade local people to plant trees in the ravaged countryside around the park. They have improved health in some villages, demonstrated alternative ways to feed cattle and fuel village fires, done all they could to spread the gospel of conservation. But the odds they face are formidable.

A first-time visitor could easily be deceived. Shimmering peacocks still danced in and out of the scattered ruins, trying to impress perpetually inattentive peahens. Monkeys still perched in the tallest trees, on the lookout for predators. And in the evenings the three lakes around which I once watched tigers prowl in broad daylight still seemed packed with protein—scores of shaggy sambar, hundreds of spotted deer, and big sounders of wild boar. I counted eighty snuffling piglets around one of the lakes, rushing back and forth in furious imitation of their elders. 'Lovely sight,' said Fateh. 'But bad sign. These are easy meals for tigers. Too many piglets are an indicator that predation is nearing zero.'

There were other bad signs too. No patrols were seen anywhere. Along upland roads closed to ordinary tourists because of 'repairs' nowhere in evidence, the once rich meadows through which I had watched tigers stalk their prey were stubble, chewed over by hundreds of domestic cattle whose droppings lay everywhere. Defiant herdsmen from a nearby village had commandeered a guard post on the western edge of the park originally built to keep them out and had scrawled their names on its walls in Devanagari script nearly a foot high—BHARAT RAM, CHANDRA MINA, DHAN RAJ, B.L. MINA, CHOTA

SINGH—just to taunt the Forest Department.

The Ranthambhore tigers' drive to reproduce remained strong—three tigresses had borne litters in the past eighteen months or so—but the big male that is presumed to have sired them all died suddenly last May, after an injury to one of his shoulders turned septic. The Forest Department now admits there may be as few as nine adult tigers within the park—fewer than are said to have been living there when Fateh arrived some thirty years ago.

Since my visit, there has been one hopeful sign. During the monsoon, when even the sere Rajasthan jungle turns lush and green and graziers traditionally drive their herds into the park to feed, a reinvigorated forest staff, armed only with bamboo staves, managed to fend off the graziers for the first time in years.

But the line between survival and extinction remains precariously thin. Even if the annual invasion of livestock can be permanently curtailed, if the poaching threat really has eased, and if other vigorous males remain, two or three more fatal accidents could still spell the end for the tigers of Ranthambhore.

I asked Fateh what he thought would happen to the City of Tigers if the tiger vanished from the park altogether. 'Maybe,' he sighed, 'they could call it the City of Peacocks.'

The conservation community was initially stunned by the poaching emergency in India. And there was soon evidence of poaching for the bone trade from Indochina and the Russian Far East as well. Suddenly every gain seemed in danger of being wiped out. Articles began to appear in the press declaring the tiger doomed.

'Rather than a cozy feeling from tiger land,' remembers John Seidensticker, curator of mammals at the Smithsonian's National Zoological Park in Washington, D.C., 'we realized the tiger was once again in crisis.' A big, plain-spoken man, Seidensticker witnessed the passing of the Javan tiger as a young field researcher and has never gotten over it. 'It was like mourning a death in the family,' he says. 'My first reaction was to lash out in anger. Since then I've learned that anger alone is a waste of time. We need to learn from these tragedies so they won't be repeated.'

Those lessons took time to learn. The crisis initially produced denials, recriminations, quarrels over funds among conservation organizations, bickering between advocates of captive breeding and those determined, against all odds, to save the species in the wild. And any number of rescue schemes were put forward, including a suggestion from an overzealous Briton, who sought funds with which to tranquilize and radio-collar every single tiger in the Project Tiger reserves so that a satellite could keep track of them all from space.

But real progress was made too. Representatives of most of the fourteen tiger-range countries met at New Delhi in 1994 and promised for the first time to cooperate in combating the tiger trade. The next year the Exxon Corporation pledged more than a million dollars annually for a five-year worldwide 'Save the Tiger Fund' campaign to be administered by the US-based National Fish and Wildlife Foundation. Pressure from foreign governments, including the US, helped persuade China and Taiwan to enforce their bans on the trade in tiger bones, and, perhaps in part as a result, the incidence of tiger poaching in both India and Russia appears to have fallen off since 1995. Belinda Wright and Ashok Kumar are not reassured, pointing out that the market for medicines that at least claim to contain tiger derivatives has not shrunk. They suggest that Indian traders have simply become more crafty in concealing their bloody work; certainly the illegal slaughter of some tigers as well as leopards and other wildlife in India continues. And so does the official lack of commitment that allows it to go on.

Perhaps the most hopeful sign for the tiger is that serious science is at last being enlisted in its conservation across much of its range.

It was early morning in Nagarahole National Park in the southern Indian state of Karnataka. Two massive gaur bulls occupied a roadside clearing. Gaur are the largest wild cattle on earth. The elder hadn't yet risen from his night's sleep, but he weighed about a ton and looked, just lying there, like a dark brown mountain. In what seemed to me an unwise effort to intimidate him, his young rival began a slow-motion strut across the clearing on his white-stockinged hoofs, carefully keeping in profile so he'd seem still bigger than he already was. He came to a halt beside a sizeable bush and slowly roiled its leaves with his horns. The older bull seemed unimpressed. But he lumbered to his feet and, moving at a yet more stately pace, approached a termite mound, lowered his horns, and, as slowly and deliberately as possible, nudged it over. The younger bull waited until the dust cleared, then started gliding towards a termite mound of his own.

'This will go on all morning,' Ullas Karanth whispered, reaching for the key to start our jeep. He has been coming to Nagarahole for more than thirty years, the last eleven as a field biologist working for the Wildlife Conservation Society headquartered at the Bronx Zoo. 'We can come back. They're not going anywhere.'

But they were. Just as the motor started, a tiger roared deep inside the jungle. In an instant the bulls forgot their rivalry and vanished into the undergrowth; each of them represented a 2000-pound breakfast to a tiger.

Tigers are so beautiful, so powerful, so secretive, so shrouded in myth that the daily reality of their lives can seem prosaic. Tigers do not 'roam' the forest as romantic writers like to have them do; instead they doggedly work carefully delineated territories, on the lookout for their next meal—and on the alert for any other predator that threatens access to it. They need meat, massive amounts of it, just to stay alive. An adult Bengal tigress on her own eats an average of 13 pounds every day, 4700 pounds every year; a tigress with two cubs can demand more than 6800. That's somewhere between forty and seventy kills annually. It takes enormous energy to do all that killing—observers at Ranthambhore estimated that the average tiger there made ten attempts before it managed to pull anything down—and it is therefore far more efficient for tigers to hunt big animals. The tigers of Nagarahole routinely feast on gaur, favouring them over the much smaller spotted deer that are staples of the tigers' diet in other forests.

Karanth and his colleague, Mel Sunquist of the University of

Florida, have shown that Nagarahole provides 32,385 pounds of meat for every square mile, such a staggering amount of prey in so many different sizes that three of India's large carnivores—tiger, leopard, and wild dog—are able to flourish here side by side. But in parts of Thailand, for example, where large ungulates have been hunted almost to extinction, tigers now struggle to survive on porcupines, monkeys, and 40-pound muntjacs. 'Indochina, with its huge forests, ought to provide the greatest potential for tiger preservation,' Karanth says. 'But the prey is all buggered-up.'

Studies by Karanth, Sunquist, Seidensticker, and others suggest that density of suitable prey is the most reliable indicator of how a tiger population is likely to fare. And history bears them out. Tiger hunting and loss of habitat were once blamed for the loss of the Bali, Caspian, and Javan tigers—and both surely played a part in their decline. But the latest research suggests that it was the loss of their prey that finally made their lives literally insupportable.

Karanth doesn't minimize the seriousness of poaching. 'If it continued at the levels it reached in the early nineties, it could provide the *coup de grâce* for the species in India,' he says, 'and we should be ruthless in dealing with anyone involved in it.' But there will probably always be some hunting, he continues, and a healthy tiger population can tolerate a reasonable amount of it: 'Suppose a forest holds twenty-four breeding females and each year eight of them give birth to a litter of three cubs. That's twenty-four cubs. In the natural course of things, half the cubs will die before their first birthday. If properly protected and fed, each of the surviving twelve will either disperse or kill and replace an already existing tiger. So in a healthy community there's always a doomed surplus.'

As long as poachers don't remove more than that number, Karanth argues, the population should remain more or less stable. (If poachers do exceed that number—as they may have done at Ranthambhore—all bets are off.) But if the prey base collapses, if tigresses begin to have trouble feeding themselves let alone their cubs, populations plummet.

'That sort of basic knowledge is essential if tigers are to be managed properly,' Karanth says. 'The role of good science is to set the standard for what potentially can happen in a park and then offer ways by which managers can accurately assess their own work. Instead of indulging in this crazy numbers game of trying to count every single tiger, we are trying to introduce basic sampling techniques to monitor success or failure. Studying prey—counting deer pellets and picking apart tiger scat to see what they've been eating—is unglamorous,' he admits. 'But after providing protection, management's most important role should be to build up the prey base. With enough to eat, enough space, and enough protection, tigers will take care of themselves.'

The three subspecies that survive elsewhere in Asia—Indochinese, Sumatran, and Siberian, or Amur, tigers—face all manner of threats. But the most serious in every case is the potential loss of prey.

Tigers remain in six Indochinese countries—Cambodia, Laos, Malaysia, Myanmar (Burma), Vietnam, and Thailand. But all have suffered war or civil unrest in the past half-century, and their interiors have largely been off-limits to scientists. While forest still covers much of the region, no one knows how much wildlife survives beneath its canopy.

George Schaller, the peripatetic director for science at the Wildlife Conservation Society whose pioneering 1967 study, *The Deer and the Tiger*, set the standard for scientific research on the animal, is not optimistic. 'I'm afraid there are very few tigers left in Indochina,' he says, 'and there are very few researchers to study them. Forests look intact from the air, but many are alarmingly empty on the ground. In Laos you can walk for weeks in the rain forest and never see a pugmark—less because of poaching than because local people have snared and eaten all but a few barking deer. The remaining tigers are forced to wander for miles in search of their next meal. When you ask villagers if there are tigers nearby, they answer, "Yes, one came by here a year ago." Most populations are too small and scattered to survive much longer.'

Schaller's colleague Alan Rabinowitz, one of the few scientists to have worked extensively in the region, agrees. 'Deer are disappearing everywhere,' he says. Four of the six species that were staples of the tigers' diet in Thailand have been virtually annihilated. Commerce in wild animal parts continues to flourish. All of the six range countries except Laos have signed the CITES agreement barring international trade in endangered species. But tiger parts are still sold openly in the marketplace alongside bits and pieces of the animals they once fed upon; in one Burmese bazaar tiger skin was recently being sold for $5 per square inch; an inch of rib cost $4.50.

'If the tiger is to survive in Indochina, governments will have to act fast,' Rabinowitz says. 'Local people will not save tigers on their own. Why should they? There's nothing wrong with long-term schemes aimed at involving local communities in conservation. But we haven't time to wait for them to work. We need first to find out where the remaining tigers are. Then we need a triage system like the one used in battlefield hospitals, to separate those populations large and strong enough to have some hope of survival from those probably too weak to make it. Finally, governments will have to designate protected areas for tigers, then commit the resources necessary to guard and manage them without compromise. Otherwise, we'll lose the Indochinese tiger.'

The news from Indonesia is more encouraging. It was once home to the Bali, Javan, and Sumatran subspecies. Now only the tigers of Sumatra remain, and until recently many authorities believed they were about to disappear as well. But findings by a mostly Indonesian team, headed by Ron Tilson, director of conservation at the Minnesota Zoo, suggest that reports of the Sumatran tiger's imminent demise may have been premature. The study area, Way Kambas National Park near the island's southern end, seems an unlikely source of hope. More than half a million people live along its border, and much of the forest has been logged in the recent past, some parts of it more than once. It was thought until 1995 that no more than twenty-four tigers survived within the whole park, and during the Tilson team's initial fifteen months of studying one 62-square-mile section, its members glimpsed individual tigers just twice.

But when they used modern methods of counting, including camera traps tripped by passing animals, they discovered that their study area alone—just over one-eighth of the park—was home to six tigers and regularly visited by twelve more. They now believe Way Kambas may contain as many as thirty-six tigers and are training 'rapid assessment teams' to survey Sumatra's other parks and unprotected forests to see if tigers are underreported there as well. Indonesian authorities estimate that there could be as many as 500 tigers scattered in reserves all across the island plus another 100 in unprotected areas, and Tilson and his researchers have helped the government draw up a comprehensive management plan to save as many of them as possible. 'There's an old Malay saying that attests to the persistence of the tiger's spirit,' Ron Tilson says. '"The tiger dies, but his stripes remain." In Sumatra our job is to provide the information and the means to help the people of Indonesia ensure that both the stripes *and* the tiger survive.'

The Siberian tiger once occupied Manchuria and Korea as well as the Russian Far East. Except for perhaps twenty scattered individuals thought to survive in north-eastern China and North Korea, it is now confined to a single 625-mile-long strip of mountainous terrain along Russia's easternmost fringe. When the Hornocker Wildlife Institute launched its Siberian Tiger Project fieldwork in and around the Sikhote-Alin Biosphere Reserve in 1992, the tiger's prospects looked very nearly hopeless. A series of hard winters in the mid-1980s followed by still harder economic times that accompanied the collapse of the Soviet Union drove local people, hunting for food, to take an ominous toll on the elk and deer and wild boar upon which the remaining tigers depended. Unregulated logging and mining threatened to shrink the tiger's home. Tiger poaching was rampant. Between 1992 and 1994, forty to sixty tigers were trapped or shot each year and their bones and skins sold in China.

But in the past three years the situation seems to have improved dramatically. In 1995 Russian Prime Minister Viktor Chernomyrdin called for a national conservation strategy. Patrolling was intensified, the Chinese border was better regulated, and poaching was reduced.

During the winter of 1995–96 some 650 men, led by Evgeny Matyushkin of Moscow University, coordinated by American researcher Dale Miquelle, and funded largely by the United States Agency for International Development, undertook a systematic census of the entire region. Nothing so precise had ever been attempted. Tiger tracks were followed, measured, and catalogued over 60,000 square miles of snowy mountain forest. The results surprised nearly everyone—there were signs of somewhere between 430 and 470 adult tigers and cubs, nearly twice as many animals as some had estimated just a few years earlier.

The Siberian tiger seems to be slowly edging its way back from oblivion. To continue this hopeful trend, the Hornocker Institute, working closely with Russian scientists, has drawn up a master habitat protection plan aimed at saving what remains of the tiger's beleaguered home. It calls for an inviolate core, a network of protected areas linked by corridors to allow safe dispersal of young tigers, along with careful management of the surrounding unprotected forests to ensure that logging and mining and road building do the least possible damage to tigers and their prey.

'In the Russian Far East we remain optimistic,' says Dale Miquelle. 'Yes, there's still poaching. Yes, there's a lot of logging. Yes, there's too much hunting of ungulates. But there's still a big stretch of more or less intact forest. Human pressure is low—and not likely to rise. If the Russians extract timber at a sustainable rate, if hunters can be persuaded to remove prey at a rate that allows tigers as well as themselves to eat, if the need or desire to poach tigers can be eliminated, tigers will survive in Russia for the foreseeable future.'

Setting forests aside for tigers is one thing, ensuring that they remain protected is something else again. No one disagrees with John Seidensticker's view: 'Tigers won't ultimately be safe until they're worth more alive than dead.' But that is a tall order in countries where space is at a premium and millions of people are in need of the food and fuel forests have always provided. To meet that challenge in India, the Global Environment Facility and the World Bank are supporting a 67-million-dollar ecodevelopment scheme aimed at relieving the human pressure on five tiger reserves. And last summer, Tiger Link, a new, all-India network of individuals and organizations, persuaded 320 members of parliament representing more than 250 million people to sign an appeal to the prime minister demanding that the central government reorganize and strengthen tiger protection.

More modest projects are also under way. Villagers were encouraged to reclaim and replant more than six square miles of degraded forest on the edge of Royal Chitwan National Park in Nepal and then allowed to keep half the proceeds from tourists eager to view wildlife. In the first year alone they earned $308,000 from entrance fees. Best of all from the wildlife point of view, one resident male tiger, a female tiger with cubs, and two transient males now use the area, and twelve rhinos have given birth within its precincts. Eric Dinerstein of the World Wildlife Fund, who helped guide the project, is delighted. 'It helps ensure the survival of the park,' he says, 'plus it adds to the area under protection. If we don't add more forest whenever we can, we'll end up like curators in a small museum, endlessly cataloguing our old collections rather than building new ones.' Benefits like these, derived directly from wildlife, seem to offer real hope for the future, though each park and each range country will require its own distinctive solution.

Meanwhile, at Chitwan and everywhere else where tigers survive, protection requires strict policing. 'It always will,' says Ullas Karanth. 'There is a criminal element even in the most sophisticated cities. We must deal with it in just the same way.'

Nowhere is policing more strict than at Kaziranga National Park in the eastern Indian state of Assam. 'Only God can keep people from killing tigers in other parks,' said Bhupen Talukdar, one of three range officers in charge of its antipoaching effort. He is a fierce, bearded man with bright silver rings on all his fingers. 'Here *we* do it.'

They do, indeed, though tigers are only the unwitting beneficiaries of the officers' primary concern: protecting the Indian one-horned rhinoceros. A long, spongy floodplain of the Brahmaputra River, Kaziranga shelters more than 1200 of these massive, myopic beasts—

more than half of all the wild Indian rhinos left on earth. Like tiger bone, rhino horn is used in traditional Chinese medicine, and a single horn can bring more than $8,000 on the black market—many times the average annual income of the people who live around the park. Between 1989 and 1993, 266 rhinos in India were butchered for their horns. Kaziranga lost forty-nine animals in 1992 alone.

Nevertheless, Talukdar said, 'The rhino is in the Assamese psyche. Lord Krishna is supposed to have brought it to Assam to fight an evil king, using it just like a tank. We are determined to protect it.' In 1994 he and his two equally tough-minded colleagues—Pankaj Sharma and Dharanidhar Boro—were given the job of waging the day-to-day struggle required if India is to save the species in the wild.

Kaziranga has many of the same problems that beset other sanctuaries all over Asia, plus a few distinctly its own. It is small. There is no buffer zone: villages and paddy fields march right up to its boundaries. Just across the Brahmaputra are crowded camps of poor Bangladeshi refugees, some willing for a fee to ferry poachers to and from the park at night. The state's finances are often in arrears; when I visited the park, neither the rangers nor the army of some 400 guards who work for them had been paid for weeks. The battered rifles the guards carry are no match for the automatic weapons wielded by intruders. Finally, the great river overflows its banks every monsoon, drowning hundreds of animals and driving hundreds more—tigers and rhinos and elephants included—into the nearby hills, where they are easy prey for anyone with a gun.

Yet, against all these odds, Kaziranga officials continue to hold the line: poaching within the park has been sharply curtailed. 'We are on a war footing,' says Dharanidhar Boro, 'and we are fighting wholeheartedly.' He does not exaggerate. Guards as well as poachers have been killed in the struggle to save this extraordinary place. There are some 120 permanent outposts within its borders. Within ten minutes of the sound of a shot, armed units can be on the scene. An expert rhino poacher might be able to saw off the precious horn and race out of the park again in that amount of time—though at least twenty intruders have lost their lives trying to do just that in the past four years. But tiger poaching in Kaziranga is virtually impossible. 'No one can skin a tiger in so short a time,' Talukdar explains. 'And they can't bury it either. The smell lasts for a month. It can't be hidden.'

The result is that Kaziranga remains something like a paradise for tigers and their prey. Under Ullas Karanth's direction, a team of young researchers working with camera traps in one small section of the park has recently captured enough individual animals on film for him to guess that Kaziranga as a whole may be home to more than 85 tigers, including cubs. Each cat's stripes are distinctive, Karanth explains, 'a sort of bar code by which tigers can be identified'. Prey and tiger density here may be even greater than at Nagarahole.

But Kaziranga also offers the most vivid possible reminder that the tiger is only the most charismatic actor on a crowded stage. Above the yellow ten-foot grass, against the Kodachrome blue sky, shrieking flocks of green parakeets fluttered in and out of silk-cotton trees covered with crimson flowers; the fleshy fallen blossoms, the size and shape of six-fingered gloves, patterned the path ahead of us, forming a red carpet as we rolled forward in our jeep. A crested serpent eagle struggled to get off the ground a few yards ahead of us, a big crow-pheasant pinioned in its talons.

Wherever the dense walls of grass part to reveal a clearing, there are animals in overwhelming profusion—hundreds of wild boar and hog deer and swamp deer, scores of sleek black buffalo with horns like scimitars, rhinos that look as large as fire trucks. On the edge of a shallow lake late one afternoon two groups of elephants filed gravely past one another without a sideways glance; I counted fifty-eight cows and calves and one magnificent lone tusker before they all glided into the grass again and disappeared.

The next evening a cow rhino concerned for her calf and agitated by the sound of the jeep in which Pankaj Sharma and I were riding suddenly whirled, kicking up dust, and charged straight for us. Sharma is a big man with a big voice, but when he clapped his hands and shouted to warn her off, she kept coming, amazingly fast, her broad body seeming to float above the ground, head high, ears straining to make out the source of the annoying sound—Mrs Magoo at full tilt.

We pulled away. She lost track of us, slowed, sniffed the air, and went back to grazing.

As we headed back to headquarters at dusk each evening, we passed antipoaching squads on the move along the winding forest tracks. These are the authentic heroes of conservation, little bands of two and three men wearing tattered overcoats and armed with rifles, moving through the mist. Without them this magical world would long ago have vanished. That it has not already done so despite the odds is dramatic proof that with help from scientists and support and understanding from the rest of the world Asians can save their own forests; the tiger and its world still have a future.

Back on the track of Sita in Bandhavgarh the sun was up, and somewhere high above our heads a hive of bees, awakened by its warming rays, began to hum. The elephant continued to squelch his massive way through the swamp, leaving behind footprints as big around as wastebaskets.

There were signs of tigers everywhere. Pugmarks criss-crossed the inky mud. Deep within the grass lay a clutch of whitened bones, all that remained of a chital kill.

A sleek grey-brown jungle cat, the size of one of its domestic cousins, leaped soundlessly onto a fallen tree, the better to see down into the grass. Something small was moving there. The cat arced high into the air to maximize its pounce, disappeared for a moment, then returned to the log, a mouse wriggling in its mouth, and watched us pass before settling down to eat.

The mahout nudged his elephant to the left, towards a little stream that twists along the base of the hills.

The elephant began to rumble almost imperceptibly.

A tiger was nearby.

The mahout leaned forward, peering through the undergrowth.

Then, there was Sita, sprawled out in a clump of grass overlooking the stream. The crimson rib cage of a half-eaten chital rested just a few feet away. She gazed placidly at me, just as she had eleven years before. The noisy, odd-looking burdens on the backs of elephants don't seem to register with tigers as human beings—though the sight of a man or woman walking 200 yards away would have sent her rushing off into the forest. She rolled over and was soon fast asleep again, all four paws in the air, full white belly exposed to the sky.

There was no sign of the cubs. No tiny tracks in the mud, no telltale mewing among the distant birdcalls.

Had she lost this litter, just as she had lost her last one?

After some time the mahout urged his elephant back from the tigress. He splashed across the narrow stream, then along the bottom of the hillside.

The sun was high now, the forest silent.

We started climbing, the mahout's eyes fixed on the hillside.

He stopped, smiled, and pointed upward through the leaves. It took me a moment to spot the three cubs the size of cocker spaniels lying on a little rock shelf perhaps a hundred feet above us. The two females dozed, but their brother was up and alert, his big ruffed head and his paws out of all proportion to his body. His bright eyes looked right past us, focussed on his mother far below, waiting for her signal to clamber down the hill and eat.

Here at Bandhavgarh—and in every tiger forest where there remains enough to eat and human intruders are kept at bay—the life cycle of the great cats continues. Gazing up at the cubs, I remembered something Dale Miquelle had said to me. 'We have to find the magic formula that allows man and tiger to coexist. That's not a dreamy goal. Finding it may be the key to man's survival as well. After all, we share the same ecosystem. If we can't save the most magnificent animal on earth, how can we save ourselves? I don't believe the tiger's cause is hopeless,' he continued. 'At least it's no more hopeless than our own.'

NOTES ON CONTRIBUTORS

Mahesh Rangarajan is a historian who focuses on human–wildlife interactions. He is also a popular political analyst on Indian television. He has been a Fellow of the Nehru Memorial Museum, Delhi and a visiting professor at Cornell University, USA and has been published in a number of professional journals. He has authored *Fencing the Forest* (1996) and *India's Wildlife History* (2001), and co-edited *Battles Over Nature* (2003). Rangarajan also contributes to wildlife conservation in India by being an intellectual link between social activists and wildlife conservationists.

C.E.M. Russell was a British official who began his career in Assam and Bengal before moving to Mysore in 1882 to assume charge as the Deputy Conservator of Forests, a post he held until 1896. In 1900, he published *Bullet and Shot in Indian Forest, Plain and Hill*, a rich collection of natural history anecdotes and advice to would-be hunters. Apparently a ruthless and profligate hunter, Russell confesses that he simultaneously shot three Indian cheetahs—an animal that is now tragically extinct. A forest fire-line named after Russell exists in Nagarahole National Park to this day. Curiously, he switched professions late in life, becoming a Barrister-at-Law in Madurai.

F.W. Champion was a British Forest officer who was probably the most original conservationist among them all. He served in United Provinces and pioneered the shooting of wild animals with a camera instead of a gun. His two classic books *With a Camera in Tiger Land* (1927) and *The Jungle in Sunlight and Shadow* (1933) contain examples of his fine black and white images of tigers and other jungle creatures. Champion also pioneered in India the field use of automated 'camera-traps' that employed an 'ordinary tropical model quarter plate Sanderson Camera' in combination with a special tripping mechanism and flash-lights manufactured by William Nesbit of New York. Champion's photos were published in a number of international magazines and books over the years. They retain their stark allure to this day, in a world saturated with gaudy, colourful images.

A.A. Dunbar Brander served as a forest officer in the Central Provinces (later split into the states of Madhya Pradesh and Maharashtra) for twenty-one years during the early twentieth century. At that time senior foresters like Brander continuously camped in the jungles from November until April every year. To put it in Rudyard Kipling's words, they grew wise about wild animals by living in remote forests in the company of 'uncouth rangers and hairy trackers'. Brander's careful study *Wild Animals in Central India* (1923), demonstrates his admirable emphasis on documenting the natural history of tigers in preference to telling tales about shooting them like most of his contemporaries did.

Kenneth Anderson was a Bangalore-based hunter-naturalist who wrote about several man-eating tigers and leopards that he professes to have shot. He also wrote about smaller animals and about the simple pleasures of jungle explorations that most big-game hunters did not bother to write about. Like Jim Corbett before him, Anderson's adventure tales have also been instrumental in attracting many Indian readers into the world of tigers. This editor was one such reader, who was privileged to personally know Anderson during the last couple of years of his life. Anderson's racy, and sometimes semi-fictional accounts include: *Nine Man-eaters and One Rogue* (1954), *Man-Eaters and Jungle Killers* (1957), *The Black Panther of Sivanipalli* (1959), *The Call of the Man-eater* (1961), *This is the Jungle* (1964), *The Tiger Roars* (1967), *Tales from the Indian Jungle* (1970) and *Jungles Long Ago* (1976), which was published posthumously.

William Bazé was a French adventurer who served as the 'chief elephant-catcher and tamer' for Emperor Bao Dai of Indo-China. He ruthlessly hunted many tigers in the years between the two world wars, when Indo-China's now faunally impoverished forests teemed with abundant wildlife. His natural history accounts of this region, which is relatively poorly served in hunting literature, are vividly presented in his books *Just Elephants* (1955) and *Tiger! Tiger* (1957). H. M. Burton translated Bazé's original French text into English.

Arthur Locke was a colonel in the British Army who administered the State of Trengganu in Malaya (now in Malaysia) for only two years from 1949-51. However, his brief tenure was filled with adventures because he followed up and killed several man-eating and cattle-lifting tigers, besides dealing with a growing communist insurgency. Unlike the Indian subcontinent, the southeast Asian region was poorly covered in terms of tiger hunting and natural history in the nineteenth and twentieth centuries. Consequently, Locke's *Tigers of Terengganu* (1954), despite its characteristically sparse prose, is considered a rather special volume in the annals of tiger hunting literature.

Kesri Singh was an employee of India's erstwhile royal families, first in Gwalior and later in Jaipur. He spent his career serving them either in their 'game' department or in the police. He not only hunted tigers but also helped his masters and their guests to continue to kill the big cats for 'sport' even as tiger numbers dwindled in the post-war years. Singh later appears to have rendered similar services to an American company interested in filming the pre-arranged, gruesome spectacle of a mob netting and spearing a tiger to death in Assam. Besides his shikar books, *The Tiger of Rajasthan* (1959) and *One Man and a Thousand Tigers* (1959), somewhat oddly, Singh also wrote poetry.

Jack Denton Scott was an American syndicated columnist for the

New York Herald Tribune who visited India in the 1950s to legally hunt tigers. He was responding to the Indian Government's official efforts to promote tourism through such safari hunting of tigers, even as the cat was declining towards extinction! Scott, an accomplished writer, arrived in India and tried to shoot a tiger in the forests of central India with help from a professional shikar company based at Nagpur. Scott vividly described his dubious adventures in *Forests of the Night* (1959).

A. Hoogerwerf's official engagement with the game sanctuary of Udjung Kulon in Java dates from 1937, when he was appointed the assistant to the director of the Dutch government's Botanical Gardens at Bogor. Hoogerwerf was a true conservation pioneer in that he combined natural history, scientific curiosity and a concern for species endangerment when it was not fashionable to do so. He later became the Chief of Nature Protection and Wildlife Management in Dutch Indonesia and was a member of the Netherlands Commission for International Nature Protection. In his classic monograph *Udjung Kulon: The Land of the Javan Rhinoceros* (published in 1970) Hoogerwerf recorded his observations made during 1938-1952 on the now-extinct Javan tiger.

E.P. Gee was a British tea planter who resided in Assam. He was a leading advocate of wildlife conservation in post-independence India and contributed significantly through his writings and reports to raise public and government awareness about the tragic ongoing decline of tigers and other wildlife. Gee travelled extensively through India to survey the status of wildlife. Gee also served on the Indian Board for Wildlife. His classic book *Wildlife of India* was a major force in shaping conservation attitudes and policies and led to the strong wildlife preservation laws that were subsequently enacted in the 1970s. Gee's illustrated articles on wildlife appeared in international magazines such as *Field*, *Country Life*, and *Natural History* as well as in Indian journals.

Guy Mountfort was a naturalist, adventurer and conservationist who helped to found the World Wildlife Fund, and, played a key role in the launching of the international campaign to 'Save the Tiger' in the late 1960s. He was also actively involved in persuading Asia's political leaders to establish tiger reserves and enact effective wildlife conservation laws. He was awarded the Order of the British Empire in 1970 for his efforts. Besides *Saving the Tiger* (1982), Mountfort wrote several books on his natural history experiences.

Kailash Sankhala was a senior Indian Forester of the post-independence era who gave up the gun for the camera early in his career, and campaigned assiduously from within the government system to ban all tiger hunting. He succeeded admirably in this effort, almost at the eleventh hour. Sankhala served as the Director of the Delhi Zoo, was the first Director of Project Tiger and later the Chief Wildlife Warden of Rajasthan. Sankhala remained a doughty champion of strictly protected tiger reserves free of incompatible human uses, even after his retirement. Although earlier a skeptic about the utility of modern tiger research, Sankhala changed his views in later years to become a strong supporter of advanced research.

'Billy' Arjan Singh hails from a royal family in north India. He was inspired to be a naturalist through his personal encounters in boyhood with the legendary Jim Corbett. Singh is a hunter turned conservationist who played a major role in the establishment and subsequent protection of the Dudwa tiger reserve in Uttar Pradesh. He won the J. Paul Getty Conservation award in 2005. Arjan Singh has published a number of popular books on his engagement with tigers including *Tiger Haven* (1973), *Tara: A Tigress* (1981), *Prince of Cats* (1982), *Tiger! Tiger!* (1984), *Elie and the Big Cats* (1987) and *The Legend of the Man-eater* (1993).

Valmik Thapar is a Delhi-based naturalist, writer and television personality whose single-minded preoccupation with tiger conservation issues has been internationally acknowledged. Thapar serves on several wildlife policy-making bodies of the Indian Government, and is a member of the Central Empowered Committee of the Supreme Court of India that deals with forest conservation issues. He has anchored several television documentaries including the popular BBC serial *Land of the Tiger*. Besides several books on tigers that he has either edited or co-authored over the years, Thapar has recorded his first hand observations of the wild tigers of Ranthambore in his beautifully illustrated books like *Tiger—Portrait of a Predator* (1986), *Tigers:The Secret Life* (1989) and *The Tiger's Destiny* (1992).

Vladimir Troinin, a graduate of agriculture, began his career in the 1960s as game manager in the Russian Far East. His mandate was to establish a recreational hunting facility in the temperate forests near the Amba River, where he encountered, and developed a fascination for, the Amur tigers that lived in this bitterly cold region. He later worked on whales and wrote a book about them. Troinin renewed his engagement with wild tigers and their conservation during the winter of 1986 around Vladivostock city, and wrote poignantly about that encounter in an article titled *Year of the Tiger*. This article was translated from Russian by Lise Brody and published in the University of Hawaii's journal *Manoa* in 1994.

George B. Schaller, currently the Vice-President for Science and Exploration with the New York based Wildlife Conservation Society, came to India in the 1963 to study wild tigers. The resulting book, *The Deer and the Tiger* (1967) is the first-ever scientific study of the big cat. Schaller is the world's most famous wildlife biologists and has become a legend in his own lifetime. He has inspired several generations of wildlife scientists (including this editor), through his pioneering studies of several charismatic endangered species. Schaller's major works include *The Mountain Gorilla* (1964), *The Year of the Gorilla* (1966), *The Serengeti Lion* (1972), *Golden Shadows, Flying Hooves* (1973), *Mountain Monarchs* (1977), *Stones of Silence* (1988), *The Last Panda* (1993) and *Wildlife of the Tibetan Steppe* (1998).

John Seidensticker, R. K. Lahiri, K. C. Das and **Anne Wright**

John Seidensticker is a pioneer in the radio-tracking large carnivores, having studied mountain lions in Idaho for his doctoral degree. He subsequently studied tigers in Nepal, Bangladesh and Indonesia. A senior scientist with the Smithsonian Institution's National Zoological Park, Seidensticker currently chairs the 'Save the Tiger Fund' of the National Fish and Wildlife Foundation, USA. The varied publications on tigers that he has authored, co-authored or edited include: *Great Cats* (1992), *Dangerous Animals* (1995), *Tigers* (1996) and *Riding the Tiger* (1999).

R. K. Lahiri was the State Wildlife Officer of West Bengal and **K. C. Das** was the Field Director Project Tiger Sundarbans in 1974. **Anne Wright** is a conservationist and a former member of the Indian Board for Wildlife.

Mel Sunquist and **Fiona Sunquist**

Mel Sunquist pioneered the radio-telemetry studies of several carnivores including tigers in Chitwan, Nepal. Currently a professor at the University of Florida, USA, Sunquist has also studied tigers in India and Malaysia, and, Jaguars and Ocelots in Latin America, in collaboration with his students. The editor of this volume is one such former student. He has collaborated extensively with his wife **Fiona Sunquist,** an internationally acknowledged nature writer and photographer. Their popular publications on big cats include the books *Tiger Moon* (1988) and *Wild Cats of the World* (2002) and numerous articles.

John Goodrich, Dale Miquelle, Linda Kerley and **Evgeny Smirnov**

John Goodrich, Dale Miquelle and **Linda Kerley** are American wildlife biologists who have been studying the Amur (Siberian) tigers in the Russian Far East with support from the New York-based Wildlife

Conservation Society for over a decade. They have captured, radio-collared and tracked over a dozen wild tigers and come up with innovative ways of dealing with human–tiger conflict in this harsh, cold environment. They have several scientific publications to their credit.

Evgeny Smirnov is a Russian Zoologist and tiger expert with extensive experience in the tracking of tigers in snow and collaborates with the other three authors in their long-term studies of Amur Tigers.

Geoffrey Ward is a well-known American historian, with several popular books and television serials on historical themes to his credit including *The Civil War*, *Jazz* and *Baseball*. Ward came to India in 1954 as a teenager and grew up in Delhi, developing an enduring fascination for India's people, wildlife, and conservation issues even after returning home. He has written several articles on Indian wildlife in leading American Magazines like the *National Geographic* and the *Smithsonian*. Ward has also co-authored two books on tiger-related themes, *Tiger-Wallahs* (1993) with his writer-wife Diane Raines Ward, and, *The Year of the Tiger* (1998) with photographer Michael Nichols.

COPYRIGHT ACKNOWLEDGEMENTS

This page is an extension of the copyright page.

Grateful acknowledgement is made to the following for permission to reprint extracts from copyright material:

The Random House Group Ltd for *Tiger! Tiger!* by Arjan Singh published by Jonathan Cape. Reprinted by permission of The Random House Group Ltd.

HarperCollins Publishers Ltd for *The Wild Life of India* by E.P. Gee.

Permanent Black for *India's Wildlife History* by Mahesh Rangarajan.

Rupa & Co for *This is the Jungle* by Kenneth Anderson.

Natraj Publishers for *Wild Animals in Central India* by A.A.D. Brander; *With a Camera in Tiger-land* by F.W. Champion; and *Tiger! The Story of the Indian Tiger* by Kailash Sankhala.

Penguin Books Ltd, London, for *Saving the Tiger* by Guy Mountfort.

Jaico Publishing House for *The Tiger of Rajasthan* by Kesri Singh and *Forests of the Night* by J.D. Scott.

Oxford University Press and Valmik Thapar for *A Tiger's Kingdom* by Valmik Thapar.

Geoffrey C. Ward for *Making Room for Wild Tigers*.

Vladimir Troinin for 'Year of the Tiger'.

Time for Tigers by J. Goodrich et al and *Understanding Tigers* by K. Ullas Karanth reprinted from *Wildlife Conservation®*, published by the Wildlife Conservation Society.